Working Minds

Working Minds

A Practitioner's Guide to Cognitive Task Analysis

Beth Crandall, Gary Klein, Robert R. Hoffman

A Bradford Book
The MIT Press
Cambridge, Massachusetts
London, England

MIT Press books may be purchased at special quantity discounts for business or sales promotional use. For information, please email special_sales@mitpress.mit.edu or write to Special Sales Department, The MIT Press, 55 Hayward Street, Cambridge, MA 02142.

This book was set in Stone Serif and Stone Sans on 3B2 by Asco Typesetters, Hong Kong and was printed and bound in the United States of America.

Library of Congress Cataloging-in-Publication Data

Crandall, Beth.
Working minds : a practitioner's guide to cognitive task analysis / Beth Crandall, Gary Klein, Robert R. Hoffman.
   p.   cm.
"A Bradford book."
Includes bibliographical references and index.
ISBN 10: 0-262-03351-8 (alk. paper)—ISBN 10: 0-262-53281-6 (pbk. : alk. paper)
ISBN 13: 978-0-262-03351-0 (alk. paper)—ISBN 13: 978-0-262-53281-5 (pbk : alk. paper)
1. Cognition. 2. Task analysis. I. Klein, Gary A. II. Hoffman, Robert R. III. Title.
BF311.C732 2006
152.4—dc22
                                                                           2005058030

10 9 8 7 6 5 4 3

Contents

Preface

There are three senses to the title of this book, *Working Minds*. One is the notion that Cognitive Task Analysis is the study of cognition in real-world contexts and professional practice at work. A second is the sense that practitioners of CTA are themselves engaged in the work of studying the mind. The third is the sense of studying minds when they are engaged in successful accomplishment—when things "work." Taken together, these three senses capture what this book is all about. Our reason for writing this book is to describe how to do CTA studies. Our motivation stems from our shared interest in Cognitive Task Analysis (CTA) and our shared experience with CTA methods. The chronology of our collaboration illustrates the way our interest in CTA grew over time.

In 1978, when Robert Hoffman was only two years out with his Ph.D. and still focused on the psycholingustics of figurative language, he received a request for reprints of articles from Gary Klein, who was trying to find ways to apply figurative language, in the form of analogical reasoning, to challenges such as reasoning with ill-defined goals and generating predictions in the face of uncertainty. Subsequent events and correspondence led to a realization that both researchers were following similar intellectual paths.

In the early 1980s, Hoffman began to turn from laboratory studies of sentence comprehension to applied studies, in particular the perceptual learning skills of thermographers and aerial photo interpreters. Those studies (Hoffman 1987) led Hoffman to discover how much he could learn about expert reasoning by confronting the practitioner with "tough case" scenarios.

Klein had also started to use tough cases in investigating the way firefighters made decisions (Klein, Calderwood, and Clinton-Cirocco 1986), using interviews about recent challenging incidents that revealed a great deal about expertise. At this point, however, the dialog between Hoffman and Klein centered on their interests in expertise rather than their enthusiasm for using tough cases to study expertise. In 1989

Hoffman invited Klein to participate in a small workshop on expertise (Hoffman 1992; Klein 1992). Then they collaborated on a chapter on the perceptual aspects of expertise (Klein and Hoffman 1993).

Gary Klein and Beth Crandall have worked together since 1986, when Crandall joined Klein Associates. Crandall's background in development psychology was a natural bridge to the work Klein was doing on expert-novice differences and the role of expertise in decision making. They have collaborated on many projects, co-authored articles and chapters, and co-presented workshops on Cognitive Task Analysis. They share an avid interest in research methodologies and how people's lived experience, particularly their experiences in "tough cases," can inform understanding of cognitive issues.

In 1995, while on a Fulbright in the United Kingdom, Hoffman was asked to conduct a review and analysis of all of the research that had been conducted up to that point by a company called Klein Associates. One thing that came from that review was a detailed report on the Critical Decision Method (CDM) which, it turned out, is also a method that focuses domain practitioners on the analysis of "tough cases." Not wanting that report to collect dust, Hoffman approached Beth Crandall, bugging her mercilessly about publishing a review and analysis of the CDM (eventually, Hoffman, Crandall, and Shadbolt 1998).

We have been conducting applied studies (which today we call Cognitive Task Analysis) for about twenty years. A review of our files shows that, collectively, we have done CTA projects in more than a hundred distinct domains and have conducted over a thousand CTA interviews. We have personally interviewed and observed hundreds of domain practitioners in domains including weather forecasting, clinical nursing, firefighting, nuclear engineering, mathematics, and military command and control. The roster of participants includes individuals who are indubitably world-class and at the peak of their profession, and we feel fortunate to have had opportunities to work with "the best of the best." We have also worked with individuals who are at journeyman and apprentice levels as well because a theory of cognition, and methods for studying it, must embrace phenomena that span the proficiency range.

We have each explored a variety of different CTA methods, so we have firsthand experience with many of the methods used in the field. We have also evaluated, refined, and extended our methods (repeatedly), and created new methods and new combinations of methods. Hoffman has even compared CTA methods empirically, evaluating them for their efficiencies and yield.

In addition, we have trained others to conduct CTA studies. Among the three of us we have conducted about three dozen CTA courses and workshops and trained hundreds of people. Our interest in helping others do CTA work has led us to prepare instructional, CD, and Web-based tools that we could leverage for this book. In preparing this book we have drawn on that experience in offering guidance about how to do CTA.

We want to show others how to do CTA studies because of our firm belief that CTA has a strong track record and is absolutely necessary in many modern applications of cognitive science. Researchers have relied on CTA methodology to fuel a new area in psychology—the field of naturalistic decision making (Flin et al. 1997; Hoffman in press; Klein et al. 1993; Salas and Klein 2001; Zsambok and Klein 1997). We also find an increasing demand for CTA researchers in applied communities that are developing complex systems in which teams of humans use information technologies to conduct cognitive work.

When people began asking us for a "how-to" description, we looked around and realized the need. There are a number of books that refer to CTA, but none is a true handbook; most focus more on theory than on practice. None really describe in any detail what happens in a CTA project. A budding CTA researcher would be hard pressed to go from chapters in edited volumes and actually do any particular CTA procedure. Few university programs have courses that teach much about CTA methods or methodology at either the undergraduate or graduate levels. Although there are some books and many articles reporting studies that use one CTA method or another, they contain very little information about how to do CTA. Researchers and practitioners interested in doing CTA have to learn about it someplace; this book is intended to be a starting point for anyone who wants to give it a try. This book is also intended to extend the capabilities of researchers already in the field.

In writing this book, we hope to demystify CTA by showing how it is done. Like any craft, CTA has both a formal aspect and a skill aspect. CTA is not easy. That can make it scary. "What if I try to conduct the CDM procedure and mess up?" We hope to allay such fear. "This project is handcuffed by practical constraints and roadblocks!" We hope to reduce such concern by showing how CTA researchers have pressed forward in the face of difficulty.

Also, at an emotional level, we want other researchers and practitioners to be able to enjoy and see value in this type of work. CTA is challenging, but the rewards can be both immediate and immense: helping to design new interfaces that weather forecasters will use to save lives and property, helping trainee nurses learn the heuristics that experts use to save the lives of sick infants.

We agreed from the start that the book would not be a survey of CTA methods, but instead would explain, in as much detail as we could provide, how to conduct a CTA study. Accordingly, we have written about the methods that we know best and have used the most. Some of the methods we discuss were devised by other researchers; others were developed by us. Our intent is not to promote any specific methods but to illustrate how to do CTA. It turns out that methods we have used the most are, not surprisingly, the methods that we find most useful. And these are also methods with a strong track record of successful applications. We let the results from the empirical comparisons of method efficiencies speak for themselves. However, readers are encouraged to explore other methods that we did not describe in detail. We identify a broad range of CTA methods that are available, along with criteria for selecting methods for a given project.

Another reason for writing this book is that we wanted to collaborate with each other. We believed we could learn more about CTA through this joint effort. We did. We also learned more about how to upset one another and make fun of one another. Along the way, through the debates and disagreements over the past two years, we feel we have achieved all of our goals. As we wrap up the writing by composing this preface, with our last-minute arguments fresh and our struggles with deadlines vivid, we are convinced that the quality of this book would be less if any one of the three of us had not been along for the journey.

Finally, another hope is that this is merely the first book of its kind. We do not pretend to have cast any "final words" in stone, or that everything we say here is exactly right. We expect that many scientists, both applied and academic researchers, will be keen to engage us in many rounds of debate. Cognitive Task Analysis methodology— the scientific analysis, study, and comparison of CTA methods—can be seen as a field in itself. We look forward to the extension of the CTA methods palette in the continual service of human needs.

Acknowledgments

We are very fortunate to have knowledgeable friends and colleagues who gave us such a wide range of support in writing this book.

The organizations in which we work, Klein Associates and the Florida Institute for Human and Machine Cognition (FIHMC), are full of stimulating, challenging, supportive colleagues who help each of us think well and write better. We are grateful to be able to work with such smart, talented groups of people.

The writing of the book and preparation of the manuscript were supported in part through participation in the Advanced Decision Architectures Collaborative Technology Alliance, sponsored by the US Army Research Laboratory under cooperative agreement DAAD19-01-2-0009. Thanks particularly to Mike Strub, Laurel Allender, and Larry Shattuck for helping us arrange for funding and for their ongoing support for this project.

An early draft benefited from incisive critical review from two anonymous reviewers. Their comments led us to substantially revise and improve it. Whoever you are, thank you.

Many people read portions of the book at various points in its development and gave us excellent feedback and suggestions for improving it. Thanks to Anna McHugh, Andy Mills, Terry Stanard, Karol Ross, Deb Peluso, Rob Hutton, Wil Readinger, Jenni Phillips, Cindy Dominguez, Donald Cox, and Holly Baxter.

Brian Moon worked on several early drafts of the chapter on cognitive measurement and evaluation, and the chapter here reflects his ideas and contributions to the developing field of cognitive metrics. Rob Hutton was instrumental in developing the classification concepts for knowledge elicitation methods we present in the overview of CTA methods chapter. We are grateful for their conceptual contributions.

Veronica Sanger gave us her superb organizational, graphics, and production skills. Her support, along with the editorial and proofing expertise provided by Ann Gabbard and Jennifer Shafer, helped to create a coherent manuscript out of disparate chapters.

Barb Law helped, as she always does, with administrative and contractual issues. Lisa Stevens provided wonderful support during the final stages of manuscript preparation.

Buzz Reed, who has led Klein Associates for many years, is a steady presence whose patient guidance is always helpful and appreciated.

As a leader of the FIHMC "A-team" for cognitive field research, John Coffey has been a steadfast friend, as well as a crackerjack knowledge elicitor who sees things others miss, and has thus taught us a great deal about CTA in practice. FIHMC Director Ken Ford contributed in a significant way by finding the means to support focused and extensive empirical investigations into CTA methodology.

Last, we would like to thank our fellow practitioners in the wider CTA community. The CTA community includes many researchers in the human factors, cognitive systems engineering, naturalistic decision making, and training development communities who also conduct CTA and from whom we have learned a great deal. It includes many individuals who have learned about CTA from us and whose challenging needs and questions have helped us to improve. It includes the sponsors who have supported our CTA studies and put findings to action. And it includes the domain practitioners and experts who have participated in our CTA projects and studies, shared their knowledge and skills, and allowed us to look at the world through their eyes. It has been an honor.

Working Minds

1   Introduction

The value of experience and of understanding how to apply knowledge is described in a wonderful story, a story shared so often it has become one of those enduring urban legends. The story has many different versions,[1] involving various professions and famous people. In one version of the story the scientist and inventor Nikola Tesla visited Henry Ford at his automobile factory. The factory was having some kind of difficulty with its systems, and Ford asked Tesla if he could help identify the problem area. Tesla walked up to a wall of boilerplates, scanned them briefly, and then made an "X" in chalk on one of the plates. Examination of the boilerplate showed that it was indeed faulty. Ford was impressed, and told Tesla to send an invoice. The bill arrived, for $10,000. Ford, never known for his generosity, was astonished at the cost of writing an "X" on the boilerplate, and asked for a breakdown. Tesla sent another invoice, which read:

Marking wall: $1
Knowing where to mark: $9,999

This story speaks directly to the purposes and goals of this book in two respects.

First, the story illustrates the "why" of Cognitive Task Analysis (CTA). What is it that Tesla knows, and how does he know it? What tells him what to do, with Henry Ford (not the most patient of men, by many accounts) looking over his shoulder? Capturing that knowledge and reasoning is one of the things CTA can do.

Second, the story illustrates the "how" of CTA. Cognitive Task Analysis can be thought of as a set of tools in a toolkit. Like any tool, CTA can be employed well and wisely, or it can be employed poorly or inappropriately. What tool would you use if you wanted to understand how Tesla was able to grasp the nature of the problem so quickly?

This book is about having the tools and the toolkit to understand how people think: how their minds work, what they struggle with, and how they manage to perform complex work adeptly and pluck inventive solutions out of difficult, sometimes dangerous, situations. Our purpose in writing the book is to help people learn how to do CTA—how to collect data about cognitive processes and events, how to analyze it, and how to communicate it effectively.

What CTA Offers

All CTA procedures have the general goal of helping researchers understand how cognition makes it possible for humans to get things done and then turning that understanding into aids—low or high tech—for helping people get things done better. The "work" may be that of a consumer who is using a product for the first time, or that of a weather forecaster who is trying to cope with data overload during a thunderstorm, or that of a firefighter who must figure out in seconds or minutes what to do about a dangerous situation. In all these cases, performance depends on what people know, what they perceive, what they believe, and how they think.

In many applications of CTA, the work is conducted in what are called "complex cognitive systems" (Hoffman and Woods 2000). These are work settings in which the knowledge and reasoning of individuals play a role (of course), but so do the cognition and reasoning of larger groups of people, including teams and even entire organizations. In addition, these complex cognitive systems often involve people interacting with computers and also interacting with each other via computers in intricate networks of humans and technology. Cognitive Task Analysis can show what makes the workplace work and what keeps it from working as well as it might.

Over the past several years an unusually broad population of individuals has become interested in CTA. People want to know how to do it, how to use it, and how to make it work for their organizations. Systems analysts need CTA methods to develop user specifications for new computer technologies. Trainers and instructional systems designers imagine applying CTA in order to describe the cognitive processes that need to be trained and how best to train them. Market researchers clearly understand the benefit of a lens into the minds of consumers and are discovering that CTA can offer ways to expose the thought processes involved in purchase decisions and product use. Program managers tasked with building new or improved technologies for military clients are embracing the notion that front-end analyses of the operators can help ensure that their systems work effectively. They look to CTA as a tool for understanding the cognitive requirements of those operators and the most effective combinations of humans and technology. Employers faced with a range of personnel issues wonder whether CTA could provide insights into selecting and retaining personnel. Healthcare providers and medical technology developers have begun to look to CTA to assist with enhancing patient safety and to identify and apply lessons learned from errors and accidents. Military commanders, faced with increasingly complex and dangerous missions, seek ways to best support planning and decision making in the field.

Across these many different types of work, there is recognition that CTA yields information people need. It provides leverage on deeply challenging problems, and when done well it provides solutions that can make a difference.

In writing this book, we hope to increase greatly the number of people with the skills and knowledge to conduct high-quality CTA. There are individuals and organizations with problems they cannot solve and opportunities they want to take advantage of. They need CTA tools and methods, and people who know how to apply them skillfully, across a range of problems and issues.

Unpacking Cognitive Task Analysis

A good place to begin is with some definitions.

Cognitive When the tasks that people are doing are complex, it is not enough to simply observe people's actions and behaviors—what they do. It is also important to find out how they think and what they know, how they organize and structure information, and what they seek to understand better. This is a principal reason why the word "cognitive" begins the phrase Cognitive Task Analysis. Cognitive Task Analysis is a family of methods used for studying and describing reasoning and knowledge. These studies include the activities of perceiving and attending that underlie performance of tasks, the cognitive skills and strategies needed to respond adeptly to complex situations, and the purposes, goals, and motivations for cognitive work.

Task What about the second word in CTA, the notion of a "task?" It may seem straightforward to think about "task" as people engaged in discrete activities or sequences of activities aimed at achieving some particular goal. This is a traditional notion of "task." But in complex cognitive systems, it is not always the literal action sequences—the steps—that matter as much as the fact that practitioners are trying to get things done; they are not simply performing sets of procedures. Therefore, we define task in this broader sense as the outcomes people are trying to achieve.

Analysis We use the term "analysis" deliberately. Literally, to analyze something is to break it into parts in order to understand both the component parts and their relationship in making up the whole. Cognitive Task Analysis methods provide procedures for systematic, scientific examination to support description and understanding. For scientists interested in pursuing research questions, we believe CTA presents a number of opportunities and challenges (Klein, Phillips et al. 2003). For practitioners who are interested in CTA primarily to develop tools and technologies, the "analysis" component of CTA is particularly important. Cognitive Task Analysis provides a process for systematically identifying key cognitive drivers in many types of applications.

Topics and Focus

This book is about both the "why" and the "how" of methods for studying thinking and reasoning in the course of performing real-world tasks in complex and dynamic work settings.

We primarily study adults in the workplace, and the methods described here have been developed within the world of adult work. The tasks that make up the working life of firefighters, nurses, military commanders, weather forecasters, or pilots may seem far from commonplace to you and me, but they are what fill the work lives of people in each of these occupations. Everyday tasks can also mean decisions and choices about products that face consumers on a daily basis.

Much of what we have written about here focuses on people's reports about their own, lived experience—their stories and examples and their understanding of the work they do. As we will show, CTA study can reveal the risks, time elements, opportunities, and mistakes that confront people as they work. It can help us understand the workplace: the technologies, tools, work conditions, stressors, and team interaction modes that all contribute to cognitive performance. Cognitive Task Analysis can help us consider hypothetical conditions, such as the influence of system X or technology Y, or a work practice that increases tempo by a factor of two. These are all questions that have been posed by people using CTA.

We also share some of our own stories and experiences as CTA practitioners, including what we have learned about how to apply CTA methods and how to get them to work well. What factors make the difference between a great interview and a folder full of notes that don't really say much? What techniques can determine whether data analysis yields an elegant Concept Map that conveys crucial knowledge about a domain or a bewildering mass of lines and arrows? How do seasoned CTA practitioners use a given set of methodologies to investigate problems and issues?

We have several goals in writing this book. First, the book is intended to help people learn how to do a CTA: how to collect data about cognitive work, how to analyze it, and how to use and communicate it effectively. We offer examples, guidance, and our own experiences using CTA to investigate a wide range of problems, questions, issues, and domains. Our express purpose is to give people knowledge and tools that will allow them to conduct CTA studies.

Second, the book is intended to convey the reasons to do CTA: What is CTA good for? What sorts of questions can it answer? What problems can it address?

We offer a look at how experienced practitioners apply CTA in such arenas as aviation, the military, national security, health care, firefighting, emergency response, manufacturing, nuclear power, consumer research, and many others. We hope to give readers an understanding of why CTA matters and how it can make a difference. Details of methods do not mean much if the reasons for using the methods are not clear.

We chose to write about CTA methods that we have found useful for understanding cognitive aspects of work in order to share the expertise we have gained in performing CTA and provide an insider's perspective on the process of CTA data collection. These methods are the tools we use in our own work to pursue certain types of questions and to explore the cognitive landscape. For us, core questions concern how people think and reason in complex, dynamic settings that characterize real-world tasks. That said, every method carries particular assumptions with it. Every method opens some doors and leaves others tightly shut. Knowing the strengths and limitations of a particular method is critical to using it well.

We also believe that skilled and effective use of CTA methods means understanding something about the cognitive issues they have been designed to illuminate. Our experience helping people learn to use CTA methods has left us convinced that skilled CTA practice has to combine knowledge of specific methods and techniques with some conceptual grounding in cognitive theory and research. It simply isn't sufficient to pick up a tool and place it in your toolbox. To use it well, you have to also understand why the tool was fashioned in a particular way and how the tool came to be. Doing CTA well requires knowing what a cognitive perspective can offer for understanding problems and issues of work. With that in mind, we provide an overview of current work on cognition—of how people think, reason, and make decisions—in the real world.

If we are successful in meeting these two goals, we will satisfy a third one: to expand the circle of CTA practitioners. We hope to foster a community of research practitioners who have the necessary skills and knowledge to conduct CTA, who can provide useful information and effective application to individuals and organizations, and who can advocate for its use. Although learning to do CTA can be a demanding experience, the insights and perspective to be gained make the challenges of learning to do it worth the investment. One primary reason for writing this book is to present the details of the CTA process, to provide a road map for how to conduct a CTA. We want to make the methodology more accessible and the skills involved in CTA practice more attainable.

In terms of coverage of the types of CTA methods and applications, the book is selective rather than inclusive. We chose not to write a survey volume with brief summaries of many different methods. There are some excellent survey volumes and review articles available, and we recommend that you spend some time with them in order to gain an overview of the breadth of methodologies available (Bonaceto and Burns 2003; Cooke 1994; Hoffman 1987; Jonassen, Tessmer, and Hannum 1999; Patrick 1992; Schraagen, Chipman, and Shalin 2000). Instead, we have presented an overview of the field and then homed in on a smaller number of methods to provide detailed descriptions of the CTA process, offer specific guidance, describe examples from our own work, and supply practical tips. We believe the narrower focus and specificity of detail will be particularly helpful for people who are new to CTA. We also expect that people who are experienced with other forms of behavioral task analysis, or who are seasoned interviewers, will find this book interesting and useful for expanding their skills to encompass cognitive components of performance.

Talking to the Reader

Across the pages of this book we present many suggestions. Some are pretty firm guidance about the "how to" of CTA. Some convey lessons learned or cautionary tales. Some are specific descriptions of steps in procedures and may seem rather prescriptive. Others are best regarded as advice.

In our efforts to present advice and guidance, we occasionally speak directly to you, the reader, as an individual who is interested in learning about and possibly conducting CTA. Thus, for example, we say in chapter 2:

The three primary aspects of CTA are knowledge elicitation, data analysis, and knowledge representation. Each of these aspects is critical to a successful CTA study. Many people equate CTA with the first aspect, eliciting the knowledge, because traditionally that has received the most attention. But if you don't do a good job of analyzing your data, why bother collecting them? And if you don't represent your findings so that others can understand them and why they matter, what have you accomplished?

It is our hope that this style of directly addressing the reader is not perceived as overly familiar or informal. We are simply trying to communicate clearly and in a way that is meaningful to you, the reader, as we present the concepts covered in this book.

Organization of the Book

The book is organized into three major sections:

Part I, "Tools for Exploring Cognition in Context," provides detailed guidance for planning and carrying out CTA. It includes chapters on capturing knowledge and on capturing the way people reason. We rely on this distinction throughout the book: CTA investigates what people know and how they think.

Part II, "Finding Cognition," provides a perspective on studying cognition in real-world settings and what an expanded view of cognition—a macrocognitive framework—offers. We describe some of the issues that surround CTA and what it means to study cognition in context. We end the section by exploring the challenges of rapidly changing technology.

Part III, "Putting CTA Findings to Use," describes key issues in applying CTA findings to several applications areas: technology development, training and instructional design, and market research. We also present a chapter on the role of CTA in the development of measures for evaluating cognitive work.

Our intent in writing this book is to share what we have learned about CTA, from our experience in the field to the concepts and models we draw on. We have offered examples and suggested ways to apply CTA findings to real-world problems and issues. We hope this book provides you with some tools you can use in your own practice and that the CTA methods can help you discover how people like Tesla know where to put their chalk marks.

2 Overview of Cognitive Task Analysis Methods

In this chapter, we briefly survey the leading methods for conducting Cognitive Task Analysis (CTA). The purpose of CTA is to capture the way the mind works, to capture cognition. The researcher or practitioner carrying out a CTA study is usually trying to understand and describe how the participants view the work they are doing and how they make sense of events. If they are taking effective action and managing complex circumstances well, the CTA should describe the basis for their skilled performance. If they are making mistakes, the CTA study should explain what accounts for the mistakes. Cognitive Task Analysis studies try to capture what people are thinking about, what they are paying attention to, the strategies they are using to make decisions or detect problems, what they are trying to accomplish, and what they know about the way a process works.

The three primary aspects of CTA are *knowledge elicitation*, *data analysis*, and *knowledge representation*. Each of these aspects is critical to a successful CTA study. Many people equate CTA with the first aspect, eliciting the knowledge, because traditionally that has received the most attention. But if you don't do a good job of analyzing your data, why bother collecting them? And if you don't represent your findings so that others can understand them and why they matter, what have you accomplished?

One way to get an overview of CTA is to understand how many methods there are, the sorts of labels applied to them, and what types or categories they belong to. But describing the larger picture of CTA can be quite a challenge. Cognitive Task Analysis has developed from many diverse traditions (see chapter 9) with differing root metaphors, terminologies, prevailing methodologies and testbeds, areas of application, and standards for what qualifies as worthwhile—or even what qualifies as "cognitive." In the first section of this chapter, we review methods of knowledge elicitation and present some ways to distinguish among them. In the second half of the chapter, we discuss approaches to CTA data analysis and representation.

Knowledge Elicitation Methods

Knowledge elicitation is the set of methods used to obtain information about what people know and how they know it: the judgments, strategies, knowledge, and skills that underlie performance. There are many different knowledge elicitation methods, so many that simply tracking them all down is a challenge. Tables 2.1[1] and 2.2[2] illustrate the diversity of tools and techniques available to CTA practitioners. They contain the methods that can be found at two different websites created with the express purpose of providing information about CTA. The CTA Resource site (http://www.ctaresource.com) is maintained by Aptima, Inc. All methods identified as "knowledge elicitation" within the CTA methods summary information provided at CTA Resource are presented in table 2.1. The Survey of Cognitive Engineering Methods and Uses was developed by the MITRE Corporation and can be accessed through their Mental Models website (http://mentalmodels.mitre.org/index.htm). Information provided at that website is presented in table 2.2.

Both websites provide descriptions and references for individual methods, along with a number of other resources. In addition, they each organize methods into classes or types of knowledge elicitation, and those categories and method assignments are included. However, tables 2.1 and 2.2 are by no means exhaustive. Other sources present additional methods and various ways of organizing and categorizing them (e.g., Cooke 1994; Hoffman et al. 1995; Jonassen, Tessmer, and Hannum 1999; Schraagen, Chipman, and Shalin 2000).

The first thing to notice about these two tables is the sheer number and variety of knowledge elicitation methods and tools. Even though the CTA Resource and Mental Models websites have similar goals, the methods they list and the categories they use to organize them are considerably different. Some of that difference may be due to the lack of generally accepted definitions and qualifiers for what counts as CTA in the first place.

Another reason for the diversity we see in the tables is that methods have been assimilated into the family of knowledge elicitation techniques by a number of different pathways. Some methods have been developed specifically for CTA (e.g., Goal Directed Task Analysis; PARI method); others have been purposefully adapted from methods initially created for other uses (e.g., Concept Mapping; Cloze Technique; Table Top Analysis). Still others have migrated into the field as researchers and practitioners began applying tools developed for purposes such as task analysis and instructional systems design to cognitive issues (e.g., Repertory Grid; Activity Sampling; Hierarchical Task Analysis).

Table 2.1
Knowledge elicitation categories and methods

Interview
Applied Cognitive Task Analysis
Cloze Experimental/Minimal Scenario technique
Cognitive Function Model
Comparing two or more representations
Critical Decision Method
Critical Incident Technique
Critical Retrospective
Crystal ball/stumbling block
Diagram drawing
Distinguishing goals
Dividing the domain
Focus groups/joint application development
Functional Flow Analysis
Group discussion
Group interview
Hazard and Operability Analysis
Identifying aspects of the representation
Information Flow Analysis
Interaction analysis
Interruption analysis
Job analysis
Operator Function Model
Precursor, Action, Result, Interpretation method
Questionnaires
Reclassification/goal decomposition
Retrospective/aided recall
Self critiquing/eidetic reduction
Step listing
Tabletop analysis
Teachback
Think-aloud
Twenty Questions
Workflow model

Observation
Active participation
Activity sampling
Cognitive Function Model
Controlled simulated observations
Field observations/ethnographic methods

Focused observation
Interruption analysis
Job analysis
Operator Function Model
Process Tracing/Protocol analysis
Role play
Shadowing another
Simulator/mockup
Structured observation
Time line analysis
Unstructured interview
Walk-throughs and talk-throughs

Textual
Content analysis
Management Oversight Risk Tree technique

Psychometric
Cloze Experimental/Minimal Scenario technique
Concept listing
Controlled association
Drawing closed curves
Eliciting estimations of probability and utility
Free association
Function Allocation issues and tradeoffs
Graph construction
Hierarchical sort
Laddering
Likert scale
Magnitude estimation
Multidimensional card sorting
Nonverbal reports
P Sort
Paired comparison
Q Sort
Repeated sort
Repertory Grid
Statistical modeling/Policy capturing
Step listing
Structural analysis techniques
Triad comparison

Source: www.ctaresource.com

Table 2.2

Survey of cognitive engineering methods and uses

CTA Methods	Discourse/conversation/interaction analysis
Applied Cognitive Task Analysis	Exploratory Sequential Data Analysis
Critical Decision Method	Interruption analysis
Cognitive Function Model	Minimal scenario technique
Cognitive-Oriented Task Analysis	Retrospective/aided recall
Decompose, Network and Assess method	Shadowing another
Goal-Directed Task Analysis	Shadowing self
Hierarchical Task Analysis	Simulators/mockups and microworld
Interacting cognitive subsystems	simulation
Knowledge Analysis and Documentation	Tabletop analysis
Systems	Think-aloud problem-solving/protocol
Precursor, Action, Result, Interpretation	analysis
method	Wizard of Oz technique
Skill-based CTA framework	
Task knowledge structures	*Knowledge Elicitation—Conceptual Methods*
	Cluster analysis
Knowledge Elicitation—Interview/Observation	Conceptual graph construction
Field observations/ethnographic methods	Decision analysis
Group interview	Diagramming
Questionnaires	Hierarchical sort
Step listing	Influence diagram construction
Structured interviews	Laddering
Teachback	Likert scale elicitation
Twenty Questions	Magnitude estimation
Unstructured interviews	Multidimensional scaling
	P Sort
Knowledge Elicitation—Process Tracing Methods	Q Sort
Activity sampling	Rating and sorting tasks
Cloze Experimental technique	Repertory Grid
Critical Incident Technique	Structural analysis techniques
Critiquing	
Crystal ball/stumbling block	

Source: www.mentalmodels.mitre.org

In addition, the methods identified here vary from very specific tools and techniques (e.g., Applied Cognitive Task Analysis [ACTA]; Cognitively Oriented Task Analysis) to entire classes of methodologies used across a wide range of problems in psychology and human factors (e.g., interviews, error analysis, questionnaires). None of this is wrong, but the jumble of terms and descriptive levels is certainly confusing.

Given the mixture of terms, sources, and levels it is probably not so surprising that there is no single, well-accepted taxonomy of methods available. In fact, both of the classification schemes in tables 2.1 and 2.2 make sense, but neither seems to capture fully the multiple dimensions that exist within the overall class of knowledge elicitation methods.

We have found it useful to divide knowledge elicitation tools along two separate, intersecting dimensions: how the data are collected, and where a particular method is focused.

Types of Data Collection Methods

One way of classifying CTA knowledge elicitation is by the way the data are collected—what sort of activity is involved in eliciting information? We can distinguish four ways to gather data: interviews (i.e., asking people questions), self-reports (i.e., people talk about or record their behavior and strategies), observations of performance or task behavior, and automated collection of behavioral data. Each of these activities is discussed in the sections that follow.

Interviews

The most common CTA method is a structured interview. Interview methods are widely used in CTA practice, and for good reason. Interviews are efficient—they avoid the investment of time and effort and the logistical complications that often occur with observations. Interviews can also elicit information about issues that are easily missed by the other methods. For example, if you do not conduct an observation at precisely the right moments, you might miss key dynamics or critical elements of task performance. Anticipating when those moments are likely to occur is much more easily done in laboratory settings than in real-world data collection. Moreover, in a study of naval officers (Kaempf et al. 1992), we found that virtually every incident we studied via interview hinged on some subtle issue of personality clash or lack of confidence in the skill of a cohort. These types of dynamics are rarely incorporated into simulated task scenarios and can be difficult to discern in behavioral observations.

Many CTA practitioners view interview data as extremely rich, but best treated as exploratory data and as a source of hypotheses. Findings from one interview can be treated with greater confidence when they are replicated across interviews with other participants or are corroborated by other methods.

Interviews have disadvantages as well. Many CTA methods require interviews with highly skilled professionals, and scheduling even an hour of time with busy professionals can be difficult. Moreover, getting good data depends on participants' being able and willing to reflect deeply on their performance and their work. People may be reluctant to divulge details of some events, they may be mistaken, or they may have limited information about what happened or why. Another drawback to interview methods is that many of them require well-trained interviewers. That training requires knowledge and skill that goes well beyond understanding of standard data collection and analysis procedures.

Self-Reports

A second variety of methods are based on participants generating data on their own. These methods vary from highly structured formats, such as surveys and questionnaires, to open-ended formats such as diaries and logs. Clearly, self-report formats have an efficiency advantage, because the data collection doesn't require an interviewer or skilled data collector to be present. The quality of data generated by questionnaires and rating scales obviously depends in part on the instrument itself. There is an entire scientific field and set of methodologies that surround development of scales and questionnaires that are psychometrically sound—that are valid and reliable and can be counted on to measure what they claim to measure. Simply compiling some questions and providing the list and a pencil to participants is not necessarily going to produce insights.

Questionnaires and rating scales can be valuable tools for gathering information on the concepts and items they contain. There can be advantages in knowing what sort of information you are likely to get. The disadvantage is that structured questionnaires and rating scales do not allow for the elements of discovery and exploration that are available in more open-ended reporting formats. Diaries and logs can offer those opportunities because they provide greater flexibility of format and content. However, data quality depends a lot on participants' motivation and willingness to complete entries consistently.

Finally, self-report methods assume that respondents are capable of "self CTA" and of reporting tacitly held knowledge, subtle cues and perceptions, and other cognitive elements on their own. That assumption is not backed up by research—in fact the evidence suggests quite the opposite: people have considerable difficulty reporting on their own cognitive processes (Nisbett and Wilson 1977; Wilson 2002). And as people gain experience and higher levels of skill, it becomes increasingly difficult for them to articulate the basis for their expertise and the judgments, decisions, and assessments they make so capably (Chi and Bjork 1991; Chi, Glaser, and Farr 1988; Feltovich, Ford, and Hoffman 1997; Klein and Hoffman 1993).

Observation

Observing people perform their work offers advantages and unique opportunities. If on-site observations are feasible (they often are not), we strongly advocate that the

CTA researcher take advantage of the opportunity. There are insights and types of information that it is simply not possible to get any other way. Observations provide opportunities for discovery and exploration of what the actual work demands are; what sorts of strategies skilled workers have developed for coping; how work flows across the environment, the team, and the shift; and communication and coordination issues (Roth 2002).

Observation can be particularly effective when the researchers are well trained in the phenomenon they are studying and do not require a lot of structure for their data-collection activities. Structured observation procedures, such as predetermined formats for sampling activities, may be desirable if the research demands some degree of quantification. Without an observational checklist or other predetermined format, the researchers may wind up figuring out the coding categories afterward and wrestling with category descriptions and coding instructions. They may also find uneven coverage in their data because observers were unaware of its significance. However, advance structuring can also render the observer less sensitive to what is actually going on or unable to take advantage of a rich opportunity—particularly if what is occurring is different from what was expected.

The primary disadvantage of observational methods is that they simply may not be feasible, either because the observation opportunity represents unacceptable risk to observers, or because observers get in the way and impede the ability of personnel (e.g., firefighters, medical personnel, military forces) to respond fully to a critical situation. Other issues in observational data collection are that the events observed may not be typical and that the observers have to be highly skilled in order to capture what is going on.

In our view, observation is best coupled with other forms of data collection such as interviews to find out how the participants were viewing the events. Merely recording the events and actions taken can result in a misleading or cognitively shallow account.

Automated Capture

The collection of CTA data can be handled by computers. This approach has not been widely used to date, but we expect that to change. One example is the Situation Awareness Global Assessment Technique (SAGAT) developed by Endsley (1988b; Endsley and Garland 2000). Previously, de Groot (1946/1978) had described a strategy for comparing chess players at different skill levels. The de Groot method was to have a player study a game in progress and then unexpectedly remove all the pieces. The player would be asked to reconstruct the board. De Groot found that players were more accurate when reconstructing actual board positions than they were in reconstructing randomly placed pieces and that more skilled players were more accurate in reconstructing coherent board positions than novices. The SAGAT method is an adaptation of de Groot's technique, basically a form of "time freezing." In the midst of a computer-driven simulated mission, all of the instruments go blank and the pilots are asked to reconstruct the instrument values. SAGAT is a measure of situation awareness. According to Endsley, the better a person's situation awareness, the more accurate the reconstruction.

Advantages and drawbacks to automated capture are similar to those we noted for questionnaires and surveys. Automated capture offers ease and precision of data collection. The potential naturalness of embedding data capture in the computer-guided flow of events has benefits and appeal. Disadvantages include the effort to program the system, the difficulty of determining when to interrupt task performance, and the insensitivity of the knowledge capture to nuances, confusions, and questions that the participant might raise. Another limitation is that the automated capture is not well suited for follow-up interrogation or deeper probing to follow up participants' comments. Automated capture doesn't lend itself to the back-and-forth, interactive data gathering that is possible in interview and observational settings.

Types of Data Targets: Where Are Methods Focused?

A second set of CTA categories addresses *where* to look for data, rather than how to get them. Here, we consider four different facets of the data collection target: its location along a continuum in time, in realism, in difficulty, and in generality.

The two ways of categorizing knowledge elicitation methods intersect. Table 2.3 illustrates the intersections between how to look and where to look. For example, one can conduct an interview about a retrospective event or observe a videotape of a past event; one can observe exercises and events as they are occurring in present time and interview participants as it unfolds, and so on.

Time

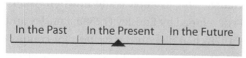

How close to "here and now" is the data target?

Studying cognitive performance, we can work with participants as they are in the midst of performing a task or working with a problem and collect data concurrently

Table 2.3

Key attributes of CTA methodology

How to Look: → / Where to Look: ↓	Interview	Self-Report	Observation	Automated Capture
Where in TIME: past/present/future				
Where in REALISM: real world/simulation or scenarios				
Where in DIFFICULTY: routine tasks/challenging tasks				
Where in GENERALITY: abstract knowledge/specific events				

in present time. We can also elicit data about events that have happened at some point previously (i.e., retrospectively). We might ask about events that are likely to happen in the future, or about hypothetical possibilities. Each of these possibilities has advantages and drawbacks.

One of the most powerful means of eliciting knowledge is to study prior incidents that were extremely challenging, to see what made them so difficult and to learn why decision makers succeeded or failed. Flanagan (1954) introduced the idea of using critical incidents to describe the nature of work, and Hoffman (1987) showed that the study of tough cases resulted in high degrees of efficiency in eliciting knowledge from experts. Klein, Calderwood, and MacGregor (1989) and Hoffman, Crandall, and Shadbolt (1998) described the use of a Critical Decision Method for knowledge elicitation that relies on retrospective accounts. We discuss CDM in detail in chapter 5.

Retrospective data can provide access to particular types of incidents. For example, we might elicit data about critical incidents of a particular type (e.g., emergency response to tornados) or to a specific event (e.g., emergency response to Hurricane Katrina). Retrospective accounts are usually studied via interview, but it is also possible to rely on self-report. A guided questionnaire could enable a person to review a prior incident and provide some description of how judgments and decisions were made. Retrospective data collection allows researchers to focus on particular types of events and aspects of cognitive performance.

The primary disadvantage of retrospective incident accounts is that people may forget or even distort key details. Memory is fragile. Therefore, data from retrospective accounts should be treated as a source of hypothesis or as a record of events that requires independent verification. For that reason, we recommend the use of converging operations and other forms of cross-checking of results of retrospective inquiries.[3]

Collecting data concurrently in time avoids many of the memory difficulties noted above (Ericsson and Simon 1984). It also allows data collectors to observe and document aspects of the situation independently of the participants' perceptions. However, concurrent data collection does not necessarily ensure better access to cognitive processes if interviewing or other types of self-report are part of the data collection process. Depending on the type of activity, the act of reporting about ourselves and our behavior can introduce biases and distortion into the data. Moreover, reporting on an activity while one is performing it can disrupt and alter the very cognition we're attempting to study (Melcher and Schooler 1996; Schooler and Engstler-Schooler 1990). The distortions and disruptions may limit the circumstances and types of tasks in which these methods can appropriately be used. A fireground commander might be willing to "think out loud" during a field exercise, but doing so during an actual event would be an unacceptable distraction.

Asking participants to report on hypothetical or imagined future events can provide interesting data when those reports are tightly linked to actual events (for example, asking participants what it would have meant if a key aspect of an incident they experienced were altered in a particular way).

Realism

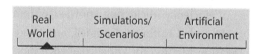

How much like the real world is the data target?

CTA data are often gathered in real-world settings. Most CTA studies are focused on performance of real-world tasks, and collecting data in real-world circumstances remains the gold standard for CTA researchers. However, there are many other types of settings that can provide CTA data that are valuable, interesting, and informative. Cognitive Task Analysis data collection is frequently carried out in simulations or in contrived or created settings. Many military field exercises contain virtually every aspect of real events save for live ammunition. Highly sophisticated flight simulators are capable of mimicking actual events—in fact, simulators have been used to "replay" incidents, including accidents and near misses, in order to better understand how they may have occurred. Similar high-fidelity simulators are now available in many fields and are used for training individuals and teams and for design, test, and evaluation of new tools and technologies. Simulations range from these very high-fidelity versions to paper-and-pencil scenarios that present key aspects of an incident and ask participants to play out action, detect problems, assess situations, and make decisions. What matters in lower-fidelity simulations is the degree of cognitive authenticity the scenario is capable of creating, regardless of its technical simplicity. Computers are being increasingly used to create gaming environments and to present humans with varieties of experience in simulated settings and artificial worlds. All of these settings offer the potential for putting humans into cognitively complex and challenging circumstances in order to understand how we perform tasks, make sense of what is going on, act, and react.

A disadvantage of using simulations is that they require a great deal of effort and expense to set up compared to going out into the field and watching people in action. Another limitation is that simulations are inherently artificial. No matter how meticulously detailed they are, the researchers will only learn about conditions that have already been tagged as important and inserted into the scenarios. Simulations are inherently constrained to a simplified version of reality. Without validation, one can never be entirely sure that the behaviors and reactions in the simulation would also occur in a natural setting. Researchers like to say that a simulation was so close to real that the participants reported being "totally wrung out" when they finished. But participants know, just as we do, that the situation isn't the real thing. Simulations do not fully capture the stress of putting lives in jeopardy or the feeling of mental exhaustion from balancing a range of difficult tasks.

Difficulty

How close to everyday events is the data target?

The tasks that we seek to understand may be highly routine, reflecting aspects of people's work environment that they encounter every day. Observational methods are useful in these instances to understand and document the full range and extent of activities that may be involved in carrying out a set of tasks. In contrast, we may want to focus data collection on tough cases and seek opportunities to observe or interview or collect reports about situations that were particularly challenging, where people's skills and knowledge were pushed to the very edge. Incident-based methods are particularly well suited for these purposes. However, tasks that are cognitively challenging are not necessarily rare events. We may want to focus on atypical or unusual occurrences in order to understand how people make sense of them and respond to them. Aviation, health care, and the nuclear power industry are three fields in which investigation of atypical events have been critically important for understanding accidents and errors and improving safety. An example is the extensive investigation that followed the nuclear accident at Three Mile Island in 1979, which helped create the field of cognitive systems engineering. Clearly, researchers may need to use different methods for studying unusual tasks. In cases where they may not have enough resources in the form of time, funding, and energy to wait for lightning to strike, researchers should use more productive and efficient methods such as simulations or retrospective techniques that do not depend on data collectors' being present for the atypical event.

Generality

Abstract/ General	Job/ Task	Incident/ Event

Is the data target to gather abstract knowledge or specific events?

Some forms of knowledge elicitation center on mapping the declarative knowledge people have in a domain. In some cases, data collection is directed at simply surveying the participant's knowledge base of factual information (which supports other cognitive processes). In other cases, the researchers examine the conceptual relationships a person has formed. One example of general knowledge capture is Concept Mapping (Novak 1991), a technique for depicting core concepts and their relationships. We discuss Concept Mapping in detail in chapter 4.

Knowledge elicitation can search for general themes within a specific job or task. Concept maps can be focused at this level. Some other methods for surveying general knowledge center on the goals people have in performing a task, and the hierarchical relationships between these goals. Annett (1996) has described a Hierarchical Task Analysis (HTA) method that elicits goal hierarchies. Goal-Directed Task Analysis (GDTA) (Endsley, Bolte, and Jones 2003) is another example of elicitation methods that focus on goals and goals structure. An advantage of many general data gathering methods is that they are fairly well structured and can be performed by researchers who have not had a great deal of experience with the techniques. The disadvantage is that they may elicit the broad, surface features of the cognitive landscape rather than the deeper layers involved in resolving competing goals or carrying out cognitive functions under complex conditions.

Illuminating these aspects of cognition requires collecting data about actual events, about specific instances where people had to make sense out of the situation, and figure out what to do and how to do it. Understanding cognition in context means understanding both the cognition and the context that surrounds it. The depth and richness of detail means that data is more fine-grained and more tightly linked to specific cues and factors, goals, settings, and people's experience. It can be a significant challenge to identify general themes and overarching meanings in data at this level of specificity.

Combination of CTA Methods

So far in this chapter, we have discussed individual methods and the various ways to classify and categorize them as separate strands. However, in many CTA projects methods are used in combination. Using various tools and techniques in conjunction provides greater leverage and deeper insight. Understanding what the various methods offer and how they can work together in various data collection settings is part of developing expertise as a CTA practitioner.

For example, interviews can be conducted while a participant is performing an actual task or a contrived task or as part of a recall of a challenging task. Interviews can also cover general knowledge that is not related to any specific incident or task. Furthermore, interviews can vary from highly structured formats to totally unstructured, "think-aloud" techniques. One might use think-aloud problem solving with test case materials derived from archived interviews. One might take the probe questions used in a particular interview technique and use them while shadowing skilled performers and conducting observations in the workplace, and so on. New methods and combinations of methods appear in the research literature all the time. Knowledge elicitation clearly does not involve an easy listing of a handful of clearly delineated methods.

Knowledge elicitation is a critical step in performing CTA, but it is only the first step. Knowing what to do with data once it is in hand is entirely as important as knowing how to get it in the first place. We turn now to a discussion of the other two elements of CTA: analysis and representation of CTA data.

Data Analysis and Knowledge Representation

The analysis phase of CTA is the process of structuring data, identifying findings, and discovering meaning. Knowledge representation includes the critical tasks of displaying data, presenting findings, and communicating meaning. Methods for analyzing

and representing CTA data have not received the same level of attention that has
been directed at knowledge elicitation. Many knowledge elicitation methods have
analysis processes and representational formats contained within an overall methodology so that the output of the elicitation process is a particular analysis product (i.e., a
representation). An obvious example is a Concept Map, which is the product associated with the elicitation and analysis process of Concept Mapping. Other examples
include the "blackboard structures" produced by COGNET (Nii 1986a, 1986b; Zachary,
Ryder, and Hicinbothom 1998) or the hierarchies produced by methods such as GOMS
(Kieras 1988) or HTA (Annett 1996; Shepherd 2000). In much of the CTA literature,
analysis and representation are inherently linked to knowledge elicitation and are not
treated as separate processes at all. Instead, distinctions between analysis and representational tools and formats are embedded in comparisons of various approaches to
knowledge elicitation (e.g., Cooke 1994; Hoffman 1987). We are aware of only a handful of articles or chapters that focus specifically on the analysis and/or representation
phases of CTA, providing examples and comparison among tools and formats (e.g.,
Hoffman, Crandall, and Shadbolt 1998; Hutton et al. 1998; Militello 2001; Wong
2004).

However, many knowledge elicitation methods produce data that can be analyzed in
many different ways and represented using a variety of formats. Treating CTA data
analysis and representation separately from knowledge elicitation allows us to see different analysis processes, products, and representation formats more clearly. We can
think about the range of possibilities available and how they might be brought together in a project to take full advantage of the CTA data.

What sorts of analysis and representation products are available to CTA practitioners? The CTA Resource website provides a catalogue of sixty different analysis tools
and approaches (presented in table 2.4). Approximately one-third of the methods identified are linked to specific knowledge elicitation methods, and many are further linked
to specific types of analysis processes and representations. The types of analytic products they yield include:

- Textual descriptions
- Tables, graphs, and illustrations
- Qualitative models, such as flowcharts, and
- Simulation, numerical, and symbolic models, including computer models.

Many of the methods identified in table 2.4 have predetermined analytic products.
But how to proceed when this isn't the case? A challenge in data analysis comes in
working with semistructured or unstructured knowledge elicitation methods. Here,

Table 2.4
Methods that produce an analytic product or representation

Applied Cognitive Task Analysis	Information Flow Analysis
ACT-R	Interaction analysis
Barrier and work safety analysis	Job analysis
Clustering routines	Laddering
COGNET	Link analysis
Cognitive Function Model	Magnitude estimation
Cognitive Work Analysis	Management Oversight Risk Tree technique
Comparing	Man–Machine Integration Design and Analysis System
Conceptual graph analysis	
Content analysis	Multidimensional card sorting
Control task analysis	Multidimensional scaling
Correlation/covariance	Network scaling
Diagram drawing	Operator Model Architecture
Discourse/conversation/interaction analysis	Operational sequence analysis
Discrete event simulation	Operational sequence diagrams
Distinguishing goals	Operator Function Model
Executive Process/Interactive Control	P Sort
Event trees	Paired comparison
Exploratory Sequential Data Analysis	Process tracing/protocol analysis
Failure models and effects analysis	Reclassification/goal decomposition
Fault trees	Repeated sort
Free association	Repertory grid
Functional Abstraction Hierarchy	SOAR
Functional Flow Analysis	Social organization and cooperation analysis
Goals, Operators, Methods, and Selection	Statistical modeling/policy capturing
Graph construction	Strategies analysis
Grounded theory	Structural analysis techniques
Hazard and Operability Analysis	Time line analysis
Hierarchical sort	Work domain analysis
Hierarchical Task Analysis	Worker competency analysis
Identifying aspects of the representation	Workflow model
Influence diagrams	

Source: www.ctaresource.com

the CTA practitioner faces the task of structuring the data, often in a series of analytic steps, to arrive at a set of findings and representational products.

There are some approaches to data analysis and representation that are useful for working with less-structured knowledge elicitation methods. Chapter 7 provides a more detailed description of different analysis and representational products.

Capsulizing Incidents

Incident-based knowledge elicitation methods, such as the Critical Decision Method, can produce voluminous data records. A single two-hour interview that is tape-recorded can run twenty to sixty transcribed pages. Even a small project can produce a lot of material to think about. One technique is to reduce incident accounts to a few pages, perhaps even to a graphic on a single page that captures the key decisions and the prominent cues. These encapsulated descriptions are easier to work with and compare than the full incident description. Narrative descriptions can also be effective as representations, because they can be created in ways that highlight cognitive content while retaining context and chronology of the event.

Cataloguing Cues and Patterns

Data records and interview notes can be examined for the cues that go into effective performance. These can be compiled by individual incident or combined across similar incidents. The cue sets can include obvious cues that novices would notice as well as subtle cues that only experts would readily detect. They can include cues that are easy to articulate as well as complex cues that require illustration. They can include relatively unitary cues as well as patterns of cues. The resulting critical cue inventory can be compiled from the notes, from transcripts, from situation awareness records, or from any other form of data. As a representation, it can convey the detail associated with specific cues, along with the pattern of a configuration of cues.

Identifying Themes

The simplest, most flexible, but most demanding approach to data analysis is to carefully review the data in search of major themes. This strategy is inductive—to work from particulars in order to discover general themes. For example, you may find the strategy for handling a particular cognitive challenge that occurs in one incident account or set of observational data being repeated in other parts of the data set, suggesting a more general finding. Key themes can be organized into a table that lists the dominant themes and cross-references them to interviews or observations or incident accounts. In this way you create an audit trail for the thematic analysis and develop the basis for additional analyses.

Coding the Data

Cue categories, thematic analysis, and other sorts of analysis products lend themselves to simple quantitative analysis because researchers can code the data and tabulate frequencies. For example, it might be interesting to create a frequency count of the typicality of themes or cue patterns: do you always see them, or rarely see them, or are they linked to certain conditions in the task or the environment?

Be advised that coding activities often lead to discovery of ambiguities that may lead back to the start and additional coding of the data, but that is a part of the learning process. The more explicit the coding rules are, the faster you will discover ambiguities. Because data coding can be so subjective, it can be important to share the task with another analyst (to seek replication of your ideas), or to turn it over to two or more coders who were not part of the effort to define the categories.

Describing Cognitive Sequences

In data that have a dynamic quality, where timing and sequence are an important part of events, data can be depicted to reflect the flow of cognitive activities of the actors. For example, sequences might be created to show the types of decisions made at various points in the incident, the cues that were present, the types of demands for identifying problems or categorizing situations, the types of strategies for gathering evidence, and so forth. Chronologies can provide temporal representations of events, specific cognitive processes, and/or cognitive requirements.

Summary

We have reviewed many different approaches to CTA in order to show the possibilities that exist and to provide a context for the methods we describe in part I. Another reason for this overview is to demonstrate that there is no single right way to do CTA. Practitioners of CTA have a wide range of choices in the strategy to use in knowledge elicitation, data analysis, and knowledge representation. Instead of worrying about following an official program, practitioners are better served by tracking the cognitive phenomenon they want to understand. Getting an insightful account of this phenomenon is far more important than preserving methodological rigor that might interfere with the investigation.

Cognitive Task Analysis research is often conducted as field studies, since it comprises the initial exploration of a cognitive process or strategy that is not well understood. We argue that it is misplaced rigor—rigor mortis, in fact—to let the choice of methods overshadow the phenomenon being studied. For field research, scientific values dictate that the methods you employ be documented and that your analyses be described in sufficient detail so that others can review your efforts and replicate your findings. You will also want to document evidence that runs counter to your hypotheses. These are all appropriate measures to increase scientific validity. In contrast, a rigid adherence to experimental control during a CTA study is an inappropriate attempt to mimic the psychology laboratory.

Therefore, we recommend that CTA researchers be prepared with a range of methods that they can use or adapt. Researchers have many choices in the strategy they use in knowledge elicitation, data analysis, and knowledge representation. The remainder of this book, in a sense, is aimed at providing enough information so that researchers can make those choices.

I Tools for Exploring Cognition in Context

3 Preparation and Framing

"How do I get started?" is a common question when people are new to Cognitive Task Analysis (CTA) practice. It may seem as though the answer to that question is obvious: go collect some data. Nevertheless, our advice is to resist the urge to jump into data collection. Experienced CTA researchers take the time to do initial groundwork. Taking the time to work through these key issues at project startup will provide important guidance for data collection, and later on for analysis of the data. In this chapter we suggest some key issues to think about and activities to carry out during the initial phase of a CTA project.

Framing the CTA Project

Framing the CTA project is the task of sharpening questions, focusing its goals, and identifying any constraints. It is a process rather than a checklist. It allows you to begin to give shape to the project and to fill in necessary detail. Some of the questions that people often pose at the start are: "What are the resource requirements?" "How much time and how much money will this take?" "Where do I find subject-matter experts?" "Who should I be talking to?" "What questions should I ask?" "What is the best method to use?"

These are important questions, and the answer to every one of them is: "It depends." Framing the CTA project helps to identify dependencies and lets you answer key questions well enough to begin to get organized and pointed in a good direction. It will help you stay on track as you move into data collection and analysis; it will help you communicate to others what you intend to do, and why; it will make planning and carrying out the project go more smoothly.

Let's take a look at why framing might matter. In preparation for an upcoming project about nursing expertise, the project team had arranged to meet with a group of nurse-managers. The purpose of the meeting was to discuss the project and to enlist their help in recruiting participants. During the discussion, one of the nurses commented, "I've heard a good NICU nurse can tell 24 hours in advance of any lab test results that a baby is about to go bad" (Klein 1998). Other nurses in the room nodded. This was apparently common knowledge. When asked how a nurse could know that a baby was about to "go bad," the reply was "nurses' intuition...they just know, based on their experience with preemies." Good nurses seem to have clear and accurate "intuitions" about infant illnesses. The key question became, "What is 'nurses' intuition'? What does it look like, cognitively and behaviorally?" The study eventually centered on nurses' perceptual judgments and assessment skills and how their expertise allows them to spot problems early. The frame for the project was about expert clinical assessment skills, but a study of critical care nurses could have been about many other issues. What if the key questions had been:

- How can we train nurses new to the unit and get them up to speed effectively?
- What are the safety issues in patient handoffs, and what happens at shift change?
- How can nurses function most effectively within the larger NICU healthcare team, particularly around the issues of information exchange and decision-making authority?

Each of these questions has a distinctly different focus. They would have led to different sets of questions in a CTA interview, different types of analyses, and possibly different kinds of deliverables at project end. The sorts of people the researchers would have wanted to interview and the types of situations that would have been of greatest interest—all of these would have been distinctly different in the three possibilities described before. The entire frame of the project, and the array of decisions to be made about how to implement it successfully, would have shifted.

When you begin to plan and prepare to do a CTA study, what do you need to think about? Here are some suggestions:

Framing Questions
- What issue or need do you plan to address?
- What will you deliver at the end of the project?
- What sorts of people can tell you about this issue?
- What aspects of expertise or types of cognition do you need to know about?
- What type(s) of situation(s) will tell you the most about the issue you are exploring?

What Issue or Need Do You Plan to Address?
Sometimes researchers identify the study topic, or perhaps a job or work function, and not much more than that before launching into CTA: "I want to understand command

and control" or "I'm interested in patient safety issues." The framing task here is to go further. Is the project concerned with skilled performance? Reorganization? Downsizing? Training design? On-the-job training? Understanding errors? Risk reduction? Introducing new technology? System development? Interface design? Envisioned worlds rather than existing tasks? New product concepts? In every case, the project will benefit by defining the core issue at a next level of descriptive detail. Here are some examples of project topics and key questions that could be addressed using CTA methods:

Military doctrine   What impact will information technology have on command-and-control decision making?

Knowledge management   Are there things that our best performers know, and skills they've developed, that aren't well documented? Is it possible to capture that expertise and make it available to others?

Accident investigation   What cognitive challenges contributed to this accident? What interventions might reduce the likelihood that similar accidents will occur?

Team training   What would allow this team to share information better and coordinate its actions more effectively?

Skilled performance   What are the key cognitive elements of this job, and how could they be incorporated into current training?

New technology insertion   How can we make this new technology most useful and effective? What aspects of performance does it support or impede?

Figure out what key question the project absolutely must answer. One way to do this is to construct a "kernel statement" that describes that question or issue in a few sentences. That kernel statement can provide critical guidance for the research team throughout the course of the project.

What Will You Deliver at the End of the Project?

The issue to be addressed here is one of outcome, and what the client expects to get from the project. By "client," we refer to the funder or sponsoring organization—whoever has the responsibility and authority to define desirable project outcomes.

Clients often have strong opinions about "what the user needs" and what the outcome of the project ought to look like. You need to know what those expectations are. Sometimes you can help the client to develop a fuller understanding about where CTA results can have impact and how they can be applied. The CTA may offer support for the client's ideas, but may sometimes contradict them. Specifying the deliverables and how the client will use them is one way to begin this dialogue. To do this well may mean spending time with the client working through the framing questions. These discussions can reveal important disconnects and differences of interpretation. For example, is there a common view of who the user of the product, technology, or service is going to be?

One way to initiate this important discussion is to talk with the client about who the stakeholders are in the project outcome. Of the various stakeholders (there are often more than one), which of those people are actually going to use the tool or application? Sometimes clients aren't sure about who the primary user is likely to be. Thus, in the design of training systems, is the user the designer, the trainer who will implement the system, or the group of people who will interact with the technology to receive training and instruction? If you are generating a set of design recommendations or storyboard concepts, the immediate user is likely to be a design team. However, the design team may not have the same view or understanding of the critical design features as the end user. The earlier you can uncover these issues the stronger the focus and clarity of your project goals will be.

Tools or technologies can be designed to address the needs of an intended user but ignore the individuals and organizations who will use the products created by the new system—that is, the customers. In fact, those who are traditionally referred to as the "end users" are often not the end users. For example, one might create a new workstation to assist weather forecasters, who would be considered end users of the system. But their job is to produce forecasts to assist aviators. The aviators are the customers for the forecasting products the system will create. It is often critical to understand customer needs in depth, and integrate those needs in the CTA research process. Moreover, different customers will have different requirements and will work under different constraints. The information and/or products they need may depend on many factors. If these are not taken into account in the CTA effort, the resulting products may have the wrong form, format, and content. In a worst case, the new tool might help its intended users but increase the workload for the user's customers.

What Sorts of People Can Tell You About the Issue?

It may seem obvious who those people should be: whoever does the job or task, uses the technology, or buys the product. However, sometimes the answer is not so clear-cut. There may be a range of types of users of a tool or technology—people who use it for different purposes, or who have varying levels of skill and experience and so interact with it very differently. There may be a chain of people involved in delivering (and receiving) training. Although it might seem like a good idea to talk with all the potential informants, it is usually not practical to do so. Resource constraints force decisions about what sorts of people, with what sorts of expertise, are likely to be most helpful.

One way to bound the problem is to specify a "target user" who is a primary recipient of the product or process. The target users are the people whose job or task will be different in some significant way—how they work, what tools they use, who they work with—as a result of the information gained in the project you are designing. It will be useful to create a description of that person, once you have identified him or her, and keep that description where you can refer to it for the duration of the project. The target user provides a critical litmus test for all kinds of choices and decisions. What part of "better interface" matters most to this user, doing this job, in this setting, working under these conditions? Does the color of the background matter to the target user? If it doesn't, why is the team spending time and money figuring out the difference between these two color tones?

What Aspects of Expertise or Types of Cognition Do You Need To Know About?

The point here is to identify the aspects of skilled performance and expertise that are most central to the issue you are exploring and the deliverable you have specified. Think ahead about the cognitive functions and subtle aspects of expertise that are going to matter most. For example:

- Is it critical to find out about the perceptual cues the expert has learned to detect?
- Do you need to determine how the expert detects anomalies and achieves a sense of what is typical?
- Do you need to find out the unique strategies or "rules of thumb" that the practitioner has created?
- Is it necessary to uncover the flow of information across a team?
- Do you need to know how team members coordinate their actions and maintain common ground?
- Do you need to learn what experts think about as they attempt to project their understanding into the possible future?

Here are some examples, drawn from several CTA projects, of the target user, the key question from his or her perspective, and the cognitive performance elements those questions point toward:

Navy Air Warfare Coordinator The project goal was to develop a shipboard-based, on-the-job training program for the Navy (Pliske et al. 2000). The target user was an experienced air warfare coordinator (AWC), showing a sailor new to the AWC position what to do and how to do it. The key questions: "What does this kid know? How fast can I get him ready? Is he good enough yet to sit in the AWC seat? Can he put anybody in danger, or is he 'okay'?"

By appreciating the key questions that confronted the AWC, the researchers identified the primary issues they had to address: assessment and diagnosis of the learner; expectancies and goals for a particular training session and for the overall training period; unpacking and articulating one's own expertise; coaching skills; and instructional strategies.

Cytotechnologists   The project goal was to document the process required to accurately scan tissue biopsies and cell samples for pathology as a first step towards developing technology supports (McDermott and Crandall 2000). The target user was a certified cytotechnologist working in a hospital or pathology laboratory, scanning and cataloguing slides for review by a pathologist. The key questions: "Is this a good tissue sample? Is there a bad (e.g., cancerous) cell in here somewhere? Where is it, and what does it tell me? Do the other cells here give me any clues about what's going on with this patient? I only have about eight minutes to read this slide, so how do I use the time I have effectively?"

To help the cytotechnologists, the researchers sought to understand these cognitive elements of performance: perceptual cues and patterns of cues; detecting questionable cells; making sense of the clinical picture; diagnostic reasoning; and managing attention and maintaining focus over the course of the shift.

Army Rangers   The project goal was to describe and document skills required for clearing buildings in urban combat settings and to incorporate those skills into training software (Phillips et al. 1998). The target user was an Army Ranger squad or platoon leader coming up to speed on features of urban combat. The key questions: "Who is in this building, and where are they? Is it safe? Is it ours? Am I in trouble? Is my squad in imminent danger? What's our best route through this space?"

The cognitive issues researchers found they needed to understand in this project included: critical perceptual cues; how to build an accurate mental model of spatial elements of the situation; sensemaking and mental simulation; knowing where to focus attention; how to manage multiple demands on attention; communication and coordination across the team; and spotting leverage points in a highly dynamic situation.

In addition to the target group, there may be other important sources of information: program managers, supervisors, designers, trainers, quality assurance personnel, and people who interact with or are part of the team in which a particular job or function occurs. For example, nurses have a lot of insight into how physicians think and work. Air traffic controllers can provide important information about pilots. Moreover, if the end product of the project is an application of technology, there may be many different users whose views, skills, experience, preferences, and needs could be considered.

What Type(s) of Situation(s) Will Tell You the Most About the Issue You Are Exploring?

The issue here is one of figuring out what sort of context is going to tell you what you need to know: where to conduct observations, what kinds of people to watch and interview, and how to focus the data collection. In some cases it might be important to know about unusual cases and crises. Another project might study what a "routine shift" is like. A third type of project might concentrate on the way a practitioner is engaged with a particular tool, technology, or type of event. The CTA format depends on what you decide about the situations to study and the data collection strategies to use.

First, what situations and settings are going to help uncover and illuminate those aspects of cognition that are of greatest interest? This may seem obvious at first—if the primary interest is in medical decision making, then arranging for data collection in a hospital is a good bet. Now imagine yourself at the hospital, ready to start collecting data. Are you there to find out about how physicians use hospital resources to arrive at a diagnosis? You may want to station yourself at the hospital lab or radiology department for a portion of your data gathering. Are you there to learn about how physicians manage workload and multiple patients? The emergency department would be interesting. How about the immediacy of life-and-death decisions made under very high time pressure? One of the intensive care units, or perhaps surgery, would offer those opportunities. Spending some time thinking about the setting will help clarify and bring into better focus what can be gained from data collected in the context of one situation versus another.

The second aspect to consider is the data collection strategies that will allow you to get the information you want. Anyone who has spent any time in a hospital will realize that some of the settings we have just considered are more accessible than others. Researchers may have an easier time gaining entry to the hospital lab than to an ICU, for example. It may seem that "shadowing" an emergency physician during a shift would be a terrific data collection strategy: you could gather observational data and then interview the doctor in between patients. Already, you are excited! As you explore this possibility with physicians and hospital administrators, it quickly becomes clear that it isn't very realistic. Patient confidentiality concerns are one barrier, and access to the emergency physicians is another. On a relatively slow night, they may have time to be interviewed between patients. But slow nights don't happen very often, so it seems more reasonable to arrange to talk with them when they come off shift. As you can see, aspects of the setting will mold the selection of methods and tools.

Bringing the Pieces Together

As you begin to consider the framing questions, you may find yourself in a bit of a swampy stretch. The questions seem to bump into each other. It is difficult to answer any one of them without knowing the answers to all of them; the answer to any one of the framing issues alters how you think about the rest. People who enjoy solving simultaneous equations have a great time with this task! When CTA researchers work with the framing questions they typically find themselves moving back and forth and working iteratively. As this iterative process unfolds, answers to the five questions begin to fit together and form a coherent frame that bounds the problem and gives the project direction.

Examples of this iterative work can be found in a number of projects in which an initial empirical effort must be conducted in order to answer the very first framing question, "What issue or need do I plan to address?" In a project on weather forecasting (Hoffman, Coffey, and Ford 2000) the goal was to demonstrate how to create a human-centered information processing system. Beyond that, there were no specifications. Thus, the process of framing had to be "short-circuited," and preliminary CTA was conducted solely for the purpose of providing focus for the first framing question. The researchers had to identify one or more leverage points for technology infusion. Initial unstructured interviews revealed that the weather forecasting organization was suffering from the loss of expertise due to the retirement of senior forecasters. Furthermore, the expertise that was captured existed only in the form of some videotaped weather briefings. Thus, it was decided that the project would focus on creating a system to capture, preserve, and share the knowledge and skills of the expert forecasters. After that had been decided, the researchers went back to work on the other framing questions, including what CTA methods to use as the best procedure for conducting knowledge elicitation and preservation.

Identifying Cognitive Challenges in Existing Data

Not all CTA projects start from scratch. Sometimes, particularly in system design projects, a lot of initial work has been accomplished already in the form of a behavioral task analysis or some other form of task-function description. There may be extensive data available based on techniques such as operational sequence diagrams (Dugger et al. 1999), operator function models (Chu, Mitchell, and Jones 1995) or Hierarchical Task Analysis (Shepherd 2000). The data may include output of computer-based programs such as Micro Saint (Laughery 1989) or IMPRINT (Kelley and Allender 1996) that have been developed to generate task or function descriptions. Similarly, instructional system design offers training developers similar tools for specifying tasks and functional elements (see Jonassen, Tessmer, and Hannum 1999).

There are many different methods and tools available for performing task-function descriptions. They differ widely in their formats and in the specificity and detail they offer. Sometimes the task-function descriptions are rudimentary, and sometimes they are detailed, extensive elaborations of task requirements. Nonetheless, they all have the benefit of providing a description of the tasks or functions that need to be performed.

If you are fortunate enough to have a task or functional description as a starting point, you can approach the set of framing questions differently. The fact that these analyses have already been performed suggests that the client has already specified the need and the deliverable (e.g., a system or configuration of systems that comprise core technologies of a ship, aircraft, or other complex equipment). In these cases, it is important to ask:

- Where do I focus resources?
- Where are the high payoff items?
- Where will CTA make the most difference?

One useful strategy is to survey the various tasks that have been identified and gauge which ones pose the greatest and most important cognitive challenges. These are the ones to focus CTA resources on. Hutton, Klinger, and Crandall (2003) have employed the concept of a "cognimeter" to describe this type of screening and prioritizing. At a much simpler level, the Task Diagram technique within ACTA (Militello et al. 1997) has the same goal: to take an initial look at a set of tasks or functions and figure out whether CTA will be helpful, and where to apply CTA most productively. The objective is to try to identify the cognitively complex tasks and the general categories of cognitive skill required for effective performance.

Whether you use one of these techniques, or one that you devise on your own, the point of conducting an initial screening is to identify the cognitively complex elements within the overall set of tasks and functions. You may find key cognitive elements of performance buried deeply within a task hierarchy. Taking a cognitive task perspective can reveal important linkages across functional units that are hidden when tasks are organized by function or system elements.

The cognimeter process devised by Hutton and his colleagues has three phases. In the initial screening step, candidates for CTA are identified by assessing task-function elements in terms of their procedural and mission-critical aspects. Segments that are evaluated as both mission-critical and nonprocedural are candidates for additional consideration and graduate to a second evaluation phase.

Once the high-payoff tasks and functions are identified, the next step is to assess the nature of the cognitive challenges each contains. Here, the point is to do an initial surface evaluation, rather than a full-scale CTA. The process will provide a roadmap to the set of tasks and functions that are most cognitively challenging and an initial look at the key aspects of cognition that comprise that challenge. You might consider functions such as decision making, sensemaking, planning and replanning, problem detection, coordination, information and attention management, or other types of functions that seem central to the project. Hutton, Klinger, and Crandall (2003) suggest that it is most important to organize the data in a summary table in order to directly compare components. This final analysis can be checked for validity by a subject-matter expert (SME) and fine-tuned. The systematic evaluation of task-function elements on the same set of key cognitive components makes comparisons across them more meaningful. It also provides a solid basis for deciding where CTA resources are best spent.

Getting Up To Speed

In preparing for a CTA study, researchers need to rapidly get up to speed and learn about the domain. Hoffman (1987) has referred to this as bootstrapping. It usually involves reading documents (e.g., books, research documents, articles, and Internet searches), a procedure that is more formally referred to as "documentation analysis." It can also include using unstructured and structured interviews with domain practitioners to become familiar with the way work is performed. General information interviews may cover a wider variety of topics including: gaining a sense of how a task, operator, or aspect of expertise fits within a more general framework; learning about the technology; learning about the organization; identifying potential experts and other informants; and locating relevant documents and literature. Every domain seems to have its own language and set of acronyms, and getting a head start on the vernacular can make a huge difference in the quality of CTA, especially in the early interviews.

Given the complexity of the systems, technology, skill sets, and domains where CTA is likely to be practiced, the sheer amount of relevant information can seem overwhelming. One of the skills that CTA practitioners develop is to learn to come up to speed quickly in a domain, to learn where to look and how to gain enough knowledge and information to make those first CTA interviews worthwhile.

One valuable means of preparation is to spend some time inside the practitioners' world. CTA researchers have ridden inside tanks, observed forecasting operations during a severe storm outbreak, flown in airborne command posts, followed doctors and nurses through intensive care units, gone to sea on aircraft carriers, and walked across offshore oil platforms. It may seem unlikely or impossible that you could gain access to these sorts of work settings. Who is going to let you on a tank? Or inside a critical care unit? Or into a weather forecasting station? We have been in all three of these settings as we have conducted our own CTA studies. If your sponsor is serious about wanting to make a difference, the sponsor should be able to secure the access you need so that you can observe how the work is done.

The perspective gained from even limited time as an "insider" is invaluable. For example, in the project on neonatal critical care described earlier in this chapter, it never occurred to the researchers that one of the stressors for NICU nurses is the auditory workload they experience. An NICU is a high-technology environment, and the array of buzzers, alarms, and beeps constantly going off is astonishing. Experienced nurses showed an equally astonishing ability to discern when the signal tones required their immediate attention. There were questions we asked during subsequent interviews, about keeping track of patient information and attention management, that we would not have considered before those observations on the unit. One CTA researcher, returning from a weeklong data collection trip on an aircraft carrier, described the crowded, noisy, and hot sleeping conditions he had found and how difficult it had been to get adequate rest. These aspects of the work environment affect cognitive function and quality of performance and were important factors to consider.

Summary

This chapter described activities that provide the initial preparation for a successful CTA project. A set of framing questions were presented that can provide important insights and guidance throughout the entire project. Answers to the framing questions will result in some ideas and initial decisions about:

- the issue or need that the research will address,
- the kind of deliverable or product that will result,

• the kinds of individuals who might be best as "target users" or "informant-collaborators,"
• the aspects of cognition or expertise that must be revealed,
• the kinds of research settings that might be most appropriate to the project goals, and
• the kinds of individuals who would be the best participants in the CTA procedures.

The chapter also provided guidance about how to work with task-function descriptions that may already exist and ways to come up to speed and learn about the domain.

4   Using Concept Maps for Knowledge Elicitation and Representation

A basic goal of Cognitive Task Analysis (CTA) is to help researchers understand how cognition makes it possible for humans to get work done. This involves capturing what practitioners know about their domain: its concepts, principles, and events. What practitioners know and believe about their domain—rightly or wrongly—is critical to their decision making. Thus, it is critical to the CTA researcher, who needs to know about what the practitioner knows.

A form of CTA that has come to be called "knowledge elicitation" is specifically aimed at helping the domain practitioner in expressing knowledge and then representing that knowledge in a way that others can understand and put to use. Knowledge modeling has applications in each of the ways in which CTA in general can be applied:

Marketing   What do consumers know or believe, and how does that affect their decisions?

Knowledge preservation   In many sociotechnical domains, wisdom walks out the door when the expert moves on or retires. The preservation of knowledge is thus a concern in many sectors of society.

Knowledge sharing   Once captured and meaningfully expressed, a knowledge representation can be used to form the core content of training programs and procedures, so that the knowledge can be reused, disseminated, and extended.

Decision aiding   The creation of any new technology to help practitioners make better decisions must to some degree and in some way be an embodiment of the concepts, principles, and procedures of the work domain. Imagine, for example, the absurdity of a weather forecasting workstation that does not allow the forecaster to draw lines representing "fronts." The technology must enable practitioners to apply their knowledge of concepts and principles and rely on their knowledge to search for meaning in the data.

Revealing skill   One of the things that domain experts know about is the procedures they use in their practice. They also know many "heuristics" or rules of thumb. Some heuristics are shared knowledge; others are ones they have created on their own. In addition, many experts also have a metacognitive awareness of their own strategies and how they manage their resources. All of these types of knowledge about processes and procedures *are* knowledge, and are thus fair game for knowledge elicitation. Once captured and meaningfully expressed, descriptions of proficient skills can be used in training, and aspiring individuals can have a better chance of achieving expertise by "standing on the shoulders" of the experts.

We can think of no CTA process or project where, in some way and to some degree, the CTA researchers did not have to elicit and then represent at least some of the domain knowledge. At one extreme, the knowledge comes from unstructured interviews or documentation analysis, and the representation is in the form of written notes or a typed document. This representation informs and supports the CTA project or helps the CTA researcher in coming to understand the domain, a process called "bootstrapping" (Hoffman 1987). At the other extreme, the knowledge elicitation process is a systematic empirical procedure and results in detailed and sometimes formal or even computable representations of knowledge.

Models of knowledge take on forms that are different from models of reasoning. Models of reasoning usually involve *process* descriptions. These can fall at a microscale at which decision making is reduced to keystroke reaction time and sequences of hypothetical basic or fundamental mental events. Examples are the tenth of a second it takes to access long-term memory or the tenth of a second it takes to shift attention. Models of reasoning can also fall at a macroscale of parallel, interacting processes such as problem recognition and sensemaking that are not necessarily decomposable into basic or sequential mental building blocks (Klein, Ross et al. 2003; also see chapter 7).

On the other hand, knowledge models express the content through which reasoning operates. Knowledge models express facts, concepts, principles, and event types that occur within the domain. As such, models must express meaningful propositions and will not take a form resembling flow diagrams of mental processes.

In recent years, a technique called Concept Mapping has been adopted by some CTA researchers as a method for both eliciting and representing knowledge. We begin this chapter by giving a capsule view of the background of Concept Mapping in research and applications. This background is important because the track record of Concept Mapping speaks to its justification and uses in CTA. Next, we present some examples of the applications of Concept Map knowledge models. We then present guidance on

Concept Mapping as a procedure for knowledge elicitation interviews, which can be conducted with individual practitioners or with small groups of domain practitioners.

Concept Mapping

Concept Maps are diagrams that are used to represent and convey knowledge. Since Concept Maps are a good means of conveying knowledge, we can use a Concept Map in figure 4.1 to express some of the ideas that we will explore in this chapter. As with any good Concept Map, figure 4.1 invites the reader to take the time to read through the propositions and understand the relations among concepts.

Background and Research Foundations

Concept Maps were developed in the course of Joseph Novak's research program in which he sought to understand and follow changes in students' knowledge of science (Novak 1977, 1998; Novak and Gowin 1984). Novak's work relied on the learning theory of David Ausubel (Ausubel 1963, 1968). The fundamental Piagetian idea is that meaningful (versus rote) learning takes place by the assimilation of new concepts and propositions into existing concepts and propositional frameworks held by the learner. This occurs by processes of subsumption (realizing how something new relates to something known), differentiation (realizing how something new draws a distinction on something known), and reconciliation (of what at first seems a contradiction of something new with something known). These terms designating learning processes will be familiar to those who have been exposed to Jean Piaget's works.

The idea of the Concept Map was created and refined over decades of work aimed at developing a method to support meaningful learning (Ausubel and Novak 1978). Today, a large research literature pertains to the use of Concept Mapping in educational settings (Cañas 1999, 2003). Indeed, Concept Maps are being used in schools and school systems around the world (Cañas et al. 1997; Ford et al. 1996). Concept Mapping as a learning exercise encourages students to use meaningful-mode learning patterns and engage in critical thinking (Mintzes, Wandersee, and Novak 2000), which results in measurable gains in knowledge and gives students an advantage that increases over time (Novak and Gowin 1984).

Building good Concept Maps leads to longer retention of knowledge and greater ability to apply knowledge in novel settings (Cañas et al. 2003; Mintzes, Wandersee, and Novak 2000; Novak 1990, 1991, 1998). Concept Mapping also has a role in evaluation.

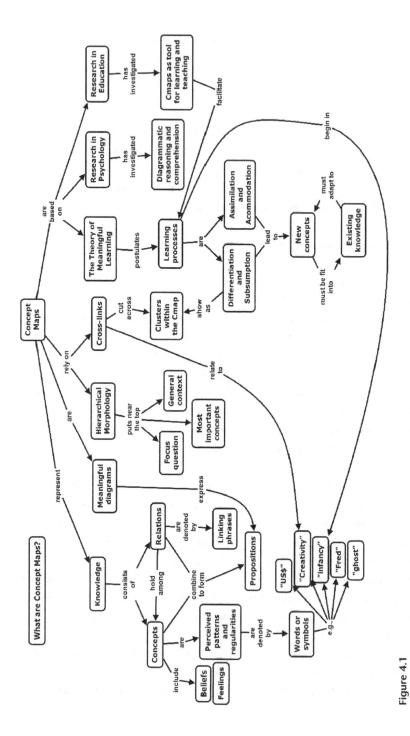

Figure 4.1

A Concept Map about Concept Maps.

It can be just as effective as more time-consuming clinical interviews for identifying the relevant knowledge a learner possesses before or after instruction (Edwards and Fraser 1983). A student's Concept Maps can help the instructor to identify knowledge gaps and the valid and invalid ideas held by the student (e.g., Markham, Mintzes, and Jones 1994).

Concept Maps have traditionally been made using paper and pencil or posters and stick-on notes. They sometimes still are (as in some business brainstorming sessions). The ability to create and share Concept Maps has been extended recently by computerization. New software tools allow students to build on their own and on one another's knowledge bases through the medium of distance learning (Chung, O'Neil, and Herl 1999).

Use of Concept Maps is not limited to primary and secondary education. Concept Mapping is being used by curriculum designers in the U.S. Navy and by university professors preparing course material, including material for distance learning (Cañas 1999). A number of companies are using Concept Mapping to preserve and share organizational knowledge, and also to provide an infrastructure for project management. NASA experts use Concept Mapping to express their views and knowledge concerning astrobiology and to come to an understanding about their definition of this new field (Cañas 1998).

Concept Maps have been used in many studies of the psychology of expertise. That work has shown, among other things, that Concept Mapping can support the formation of consensus among experts (Gordon, Schmierer, and Gill 1993). Evidence from studies of experts versus novices indicates that expertise is usually associated not just with more detailed knowledge, but with knowledge that is better organized than that of novices (e.g., Glaser 1987). Research has used comparisons of Concept Maps made by nonexperts to those made by experts to reveal expert-novice differences in knowledge organization. Concept Maps made by domain experts tend to show high levels of agreement (see Gordon 1992; Graesser and Gordon 1991). Concept Mapping has proven useful as a tool for creating knowledge-based performance support systems (Cañas et al. 1997; Dodson 1989; Dorsey et al. 1999; Ford et al. 1991; Ford et al. 1996).

Concept Maps are also being used to document communities of practice, and the conceptual, methodological, and multidisciplinary linkages that exist within a particular field. An example is shown in figure 4.2. Here a concept map is used to capture and express ideas about cognitive engineering and CTA (Hoffman, Klein, and Laughery 2002).

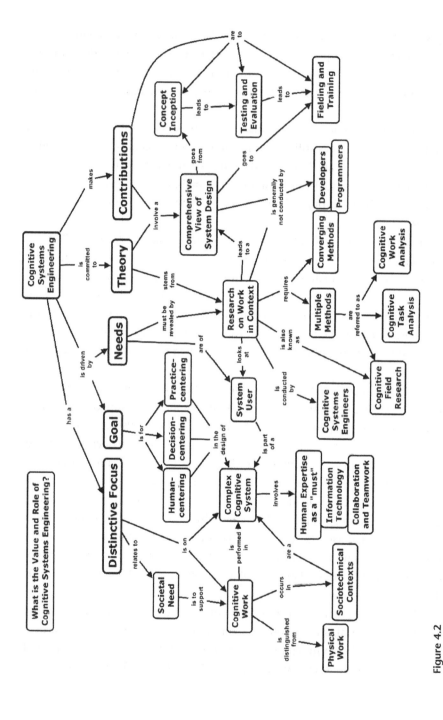

Figure 4.2

A Concept Map about cognitive engineering.

Concept Mapping as CTA

Concept Mapping is coming to be used widely as a method for eliciting and representing the knowledge of domain practitioners. In other words, Concept Mapping is now a tool in the CTA toolkit.

In order to conduct CTA in a given domain, the researcher needs to learn about the domain, that is, they need to bootstrap themselves to a sufficient level of understanding (see chapter 3). Concept Mapping can play a role here. Concept Mapping commonly triggers in the practitioner the recall of past cases in which one or more of the concepts were salient. This can be used as "incident selection" and can feed into other CTA procedures, such as the Critical Decision Method (CDM) (Hoffman, Crandall, and Shadbolt 1998).

We mentioned earlier that some of the things practitioners know about are events, processes, procedures, and their own reasoning strategies. These are fair game for Concept Mapping. In this form of diagram, referred to as a "Process Cmap," a description of an event (process, strategy, etc.) is embedded within a Concept Map. An example appears in figure 4.3. The event description is embedded in nodes and links that provide the context, or the "explanatory glue" that makes sense of the process.

Concept Maps on a given topic can be hyperlinked together. For example, the concept node "The Critical Decision Method" in figure 4.3 might be hyperlinked to another Concept Map that goes into detail about the CDM. When a set of Concept Maps is linked together this way, and organized by a "Map of Concept Maps," they form what are referred to as Concept Map knowledge models. These can be simple, consisting of a dozen or so Concept Maps, or they can be complex, consisting of hundreds of Concept Maps. The following examples show what they look like, and what they can be used for.

Example Knowledge Models

Example 1: Rocket Science The first example is from work that was performed at NASA Glenn Research Center (Coffey and Carnot 2003; Coffey, Moreman, and Dyer 1999). The motivation was the need to preserve lessons learned by retiring engineers, knowledge that would otherwise be lost to the organization. Eighteen Concept Mapping knowledge-elicitation interviews were conducted with a senior engineer, who specialized in the knowledge of a Delta rocket motor. The resulting eleven Concept Maps

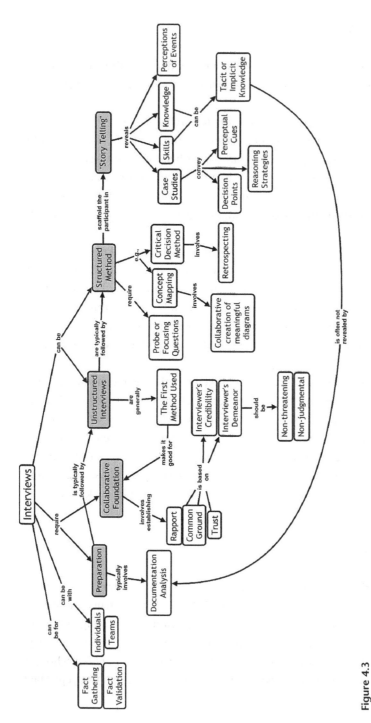

Figure 4.3

An example "CMap" on the topic of interviewing for knowledge elicitation. Process or procedural information is highlighted in gray.

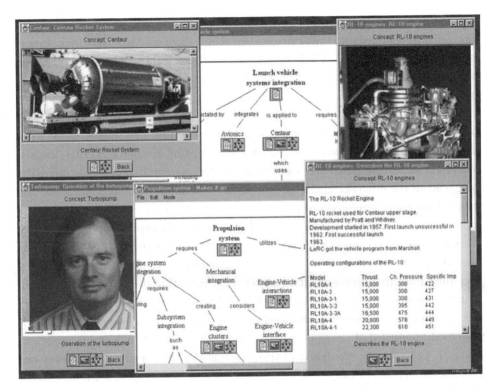

Figure 4.4
A screenshot of a Concept Map project about a NASA engineer's expert knowledge.

expressed his knowledge. Hyperlinked to the Concept Maps were 140 informative resources (diagrams, photographs, digital video interviews with the expert, etc.). A screenshot from this knowledge model is presented in figure 4.4. This includes views of two of the Concept Maps and views of some of the digital resources that were linked into the Concept Maps.

Example 2: Going to Mars   At the Center for Mars Exploration at NASA Ames Research Center, it was decided to integrate information about Mars so that it might be more effectively distributed to the public, especially to schoolchildren. At the time the project started there were many kinds of information, including books, articles, photographs, and so on. But there was no way to put the material together so that people might browse and learn from it. Over one hundred Concept Maps were created by NASA scientists to organize the material into appropriate subtopics (e.g., Mars Lander Missions,

Mars in Science Fiction, Life on Mars, etc.). Hundreds of resources (maps, photographs, URLs, etc.) were linked into the Concept Maps. These Concept Maps were then used as the indexing structure for a CD-ROM released by NASA (Briggs, Shamma, and Cañas 2001; Cañas 1999, 2003). The Concept Maps can be seen at http://cmex.coginst .uwf.edu.

Example 3: Medical Diagnostic Procedures The use of Concept Maps to create navigable explanations of decision aids was demonstrated by Ford et al. (1991). In first-generation knowledge-based decision aids (or "expert systems"), users could query the system about the inference chain that was used in reaching a diagnostic decision. The result of the query would be a sequence of formal "if-then" rules in cryptic computer code, itself not terribly explanatory. Ford et al. used Concept Mapping in twenty knowledge elicitation sessions to elicit the knowledge of an expert at first-pass nuclear magnetic resonance imaging (MRI) of ventricular function. After deriving the inference rules from the interviews, the researchers realized that the Concept Map representation of the expert's knowledge could *be* the interface for the decision aid, an interface that would support performance at a meaningful level. The resulting knowledge model consisted of ten Concept Maps and over two hundred digital resources. In using NUCES (Nuclear Cardiology Expert System), to work a given case, users can proceed through the Concept Maps that lay out the various diagnostic features. Along the way users can access resources showing representative images (clear cases, tough cases, etc.) that can be compared to those of the given case. Also included are digital videos in which the expert provides detailed discussion of various things (e.g., subtle cues the journeyman might miss). This innovation represented a milestone in the evolution of expert systems into "knowledge-based" systems. This knowledge model can be viewed at http://www.ihmc.us/users/acanas/Publications/ParticipatoryExplanation/ ParticipatoryExplanation.htm.

Example 4: Electronics Repair Procedures Another project involved capturing the knowledge and procedural skills of an expert at the maintenance and repair of a particular recording device used aboard U.S. Navy vessels. Eighteen knowledge elicitation sessions yielded eleven Concept Maps and about 140 digital resources. Again, the Concept Maps were used as the interface for a decision aid that used a question-and-answer process to guide the user through a diagnostic procedure. This system was called El-Tech, standing for Electronics Technician. A new idea that this system embodied was that it can be used in both a learning mode and a performance-support mode, that is, it is at once a training aid and a decision aid. Furthermore, it can be used by sailors at sea after being trained in the schoolhouse. This knowledge model can be viewed at http://www.ihmc.us/users/acanas/Publications/ElTech/El-Tech%20Flairs%2098.htm.

Example 5: Preserving Knowledge on a National Scale   Our final example is a project that demonstrated how Concept Maps might be used as the infrastructure for knowledge preservation on a national scale (Hoffman and Hewett 2001). In an effort to preserve traditional folk knowledge—residing in the experience of elder craftspeople in the rural villages of Thailand—a set of Concept Maps was made on the topic of Thai silk weaving. Based on transcripts of interviews with the elders, Concept Maps were made that described knowledge about each of eleven varieties of Thai silk and their methods of manufacture. These were then paired with photographs showing the silk patterns and illustrating various aspects of the weaving process. An interesting feature of the Concept Maps in this project is that a photograph of each of the various silk weaving patterns was used as the background in its Concept Map, so that one could see the pattern while reading the Concept Map about that pattern. This aspect was important for another reason—it reflected the Thai cultural aesthetic, which emphasizes color, pattern, and contrast. This Concept Map project can be viewed at http://www.ihmc.us/research/projects/ThailandKnowledgeBase.

These examples convey the sorts of knowledge models that are sought in CTA to support knowledge preservation, knowledge sharing, and the creation of decision support systems.

We can now describe the Concept Mapping procedure. First, we need to say more about what Concept Maps are, and what they are not.

What Is a Concept Map?

The idea of using diagrams to express logical statements has a rich history in mathematics, including some works by psychologist-logician Charles Peirce and mathematician Gottfried Frege. The modern idea of Concept Mapping takes this a step further, into the "user friendly" expression of meanings.

In terms of their nature as a type of diagram (or directed graph), Concept Maps involve nodes and links. The nodes represent concepts, which are enclosed in boxes. The label for most concepts is a word or just a few words, although one can also use symbols. Concepts are related to one another by meaningfully labeled linking lines. Using the link labels to express relations between two concepts, the node-link-node triads in Concept Maps form propositions, that is, they can be read as "stand-alone" simple and meaningful expressions. Example propositions in figure 4.1 are "Concept Maps represent knowledge" and "New concepts must be fit into existing knowledge." Those who are new to Concept Mapping benefit from a bit of practice at understanding the difference between propositions and sentences. For example, the single sentence, "My son plays with a red truck" has four propositions: I have at least one son, that son engages in play, that play is with a truck, that truck is red.

Concept Maps differ from other types of diagrams that utilize combinations of graphical and textual elements to represent or express meanings. For example, diagrams that Ackerman and Eden (2001) refer to as "Cognitive Maps" are large weblike diagrams with up to hundreds of "ideas" represented by the nodes. "Ideas" are typically expressed as sentences and short paragraphs. While these are a type of meaning diagram, they differ considerably from Concept Maps.

Differences between forms of meaning diagrams involve their expressiveness (semantics and syntax), their shape (or morphology), shape-meaning interactions, and dynamics. Understanding these aspects of Concept Maps is important if the goal is to create good knowledge models.

Expressiveness   In some forms of diagrams, the linking lines are unlabeled—all of the links mean the same thing. Specifically, in semantic networks and associative graphs, the lines represent a single relation: "concept X is related to concept Y." The length of the lines is sometimes used to indicate degree of associative strength or degree of semantic relatedness. Pathfinder networks (Schvaneveldt, Dearholt, and Durso 1988) and Buzan and Buzan's (1996) "Mind Maps" are such a type of meaning diagram. The links between nodes are unlabeled, and tacitly represent connections among ideas.

In Sowa's (1984) "Conceptual Graphs," the concepts can be connected using only certain kinds of logical relations such as "is a" and "has property." Concept Maps do not impose restrictions on semantics; hence their great expressive power. While logical relationships can and often do appear in Concept Maps, those relationships are regarded as just one kind of "subsumption-differentiation" relationship. Concept Maps can express many kinds of relations other than hierarchical classification, such as "explains" or "comes before."

An example might clarify the difference between classificational diagrams ("is-a" trees in graph theory), and subsumption-differentiation diagrams. A diagram might contain the following propositions:

⟨WATER is necessary for LIFE⟩

⟨WATER is necessary for EARTH'S ATMOSPHERE⟩

⟨WATER includes OXYGEN⟩

⟨OXYGEN is necessary for LIFE⟩

None of these relations involves categorization. So, while LIFE is subsumed under WATER, LIFE is not "a type of" WATER. LIFE and EARTH'S ATMOSPHERE together differentiate things for which WATER is necessary. The OXYGEN in WATER is also necessary for life, but OXYGEN is subsumed under WATER and differentiates a constituent of WATER that is necessary for LIFE, and OXYGEN is not a "type of" WATER.

Shape In "Semantic Networks," as defined by Fisher (1990), the most basic concept is located in the center of the diagram and the subordinate concepts radiate out in all directions. Concept Maps, on the other hand, are more like hierarchies in terms of their shape. There is a reason for this. In good Concept Maps, the more general or most important concepts appear toward the top and provide the context or the "big picture" for the Concept Map, while the more particular concepts tend to appear toward the bottom. One reads a Concept Map by beginning toward the top and then working downward through the levels of the Concept Map that express subsumption, differentiation, and other relations.

Shape-Meaning Interactions The shape of meaning diagrams interacts with the semantic and syntactic features. For example, in Buzan and Buzan's (1996) Mind Maps, the radiating shape combined with the impoverished semantics (unlabeled links) severely limit the diagrams' expressive power. Concept Maps are unique relative to other forms of meaning diagrams in that Concept Maps have "cross-links." Cross-links express relations that cut across the clusters or regions within a Concept Map. Examples in figure 4.1 are the propositions "Learning processes begin in infancy" and "Meaningful diagrams express propositions." Cross-links are used for a reason: in real-world domains of complexity, anything can relate to anything, and in some cases, everything does relate to many other things. Furthermore, creative insight has been defined as the result of a deliberate search for new relationships between concepts and/or propositions in one subdomain with those in another subdomain. The expression of cross-relations is facilitated by the fact that in a Concept Map one can see all of the important concepts at once.

Dynamics We note one final feature of Concept Mapping. It pertains to Concept Mapping as a process, rather than to the qualities of finished Concept Maps. Technically stated, when creating a Concept Map, the mapper uses spatiality as a tool to deconvolute meanings. As nodes and partially linked sets of nodes are moved around in the Concept Mapping space, the mapper considers various relations and ideas to be expressed. The mapper struggles to add in cross-links while at the same time avoiding the creation of a spaghetti graph having too many overlapping cross-links. Clusters of nodes will be parked somewhere, and that region of the Concept Map space becomes, in effect, a memory aid.

In one study conducted in the DARPA "Rapid Knowledge Formation" project, Concept Maps were made by domain experts but were subsequently "tidied up" overnight by computer scientists. Upon next seeing their Concept Maps so tidied up, the experts were upset because things "weren't where they were supposed to be" (Hayes, personal communication, 2003). The mappers had been using spatiality as a tool.

Concept Maps are generally referred to as representations of domain knowledge, but knowledge is itself never static, and Concept Maps are not regarded as things that are made to be cast in stone. Indeed, it is wise to always consider Concept Maps as "living" representations rather than finished "things." In capturing the expert knowledge within an organization, for instance, practitioners can always add to and modify the Concept Maps in the existing pool.

The Concept Mapping Procedure

Concept Maps can be made using pencil and paper, chalkboards, whiteboards, and even large sheets of butcher paper along with marker pens and stick-on notes. It is becoming more common for people to make Concept Maps using computers. A number of academic research groups have built software to support the creation of meaningful diagrams (Chung, Baker, and Cheak 2002; Hoeft et al. 2002). In addition, a number of commercially available software packages support the creation of meaningful diagrams. A software suite that was specifically designed to support the creation, resourcing, and sharing of Concept Maps is called CmapTools. It is available as a free download at www.ihmc.us. The Concept Maps that appear as figures in this chapter were made using CmapTools. Our discussion presumes that the Concept Mapping process is being conducted using the CmapTools software, though much of what we say is applicable if Concept Maps are made in other ways.

Concept Mapping is a skill. Some say that it encourages nonlinear thinking. It certainly does take some practice to create a Concept Map, but it takes even more practice to begin to understand what makes for a "good" Concept Map. Concept Mapping encourages—perhaps even forces—the mapper to reach for crystal clarity about what he or she wishes to express. It is important for the participating domain practitioner to be given an introductory presentation about Concept Mapping, including its research foundations. Even with such a presentation, practitioners can go into the process with skepticism. But experience suggests that some foreknowledge of what

Concept Mapping is all about, as opposed to just "diving in," can ease the interviewee into the process.

In Concept-Mapping knowledge elicitation, the researchers help the domain practitioners build up a representation of their domain knowledge, in effect merging the activities of knowledge elicitation and knowledge representation. A caution is in order, however, since the two purposes—domain knowledge representation and knowledge elicitation—can sometimes diverge, mandating differences in the ways that the two forms of interview are conducted.

During the knowledge elicitation one researcher acts as a facilitator and provides support in the form of suggestions and probe questions, while the other acts as the mapper and captures the participant's statements in the Concept Map, which is projected on a screen for all to see. The mapper needs to be proficient at quickly and accurately conducting the mapping work on the fly. This includes a facility for glancing to and from the computer monitor and the projector screen to follow the facilitator's guidance and the participant's statements.

The facilitator always walks a fine line between supporting and intruding. Recognizing the fact that this is inherently a collaborative, "co-constructive" process, to the greatest extent possible the Concept Map should express domain knowledge in the words preferred by the practitioner. However, the participant invariably benefits from assistance and suggestions. These should be couched as alternatives. So, for example, when a participant is reaching for words to express a relation that the facilitator infers is a causal relation, the facilitator might say, with rising inflection, "leads to?" "comes before?" "is a precondition for?" and the like. Typically, the practitioner will latch onto the wording that is most fitting.

The facilitator must be facile at monitoring the state of the Concept Map and how the mapper is doing while at the same time keeping track of the current discussion and making note of possible future agenda items.

It is not uncommon for a "conceptual block" to arise in a session of, say, thirty to sixty minutes. When this happens it sometimes is helpful to move to a discussion of some other part of the Concept Map or to some other topic that is to be mapped. Sometimes, when asked about the troublesome concept(s) at hand, the practitioner says something like, "Well, what I mean here is that...." Quite frequently, what the participant says captures precisely the things that should be changed in the Concept Map.

Typically, Concept-Mapping knowledge elicitation sessions last about an hour. In a one-hour session a facile mapper can expect to create about two semirefined Concept Maps, each consisting of something on the order of thirty to forty concepts and forty-five or more propositions. In successive interviews, the Concept Maps can be extended and refined.

Sometimes, the domain practitioner expresses ideas at a rate that overwhelms the mapper. It can be tempting in such circumstances to audiotape the session so that it can be transcribed and analyzed at a later time to pull out any propositions that were missed during Concept Mapping on the fly. A study of this problem showed clearly that the process of transcription and protocol analysis was so time consuming that it would have been far more efficient to recapture the "lost" knowledge by subsequent Concept Mapping interviews (Hoffman, Coffey, and Ford 2000). There are six fundamental steps to creating a Concept Map. They are described in the sections that follow.

Step 1: Select the Domain and Focus   This is a critical step in making the process and product directly pertinent to the research goals.

Concept Mapping as a CTA method must be conducted so as to tap into the practitioner's knowledge that lies at the heart of task activities. A clear, explicit focus helps to define the context and aids in the process of expressing the knowledge that is pertinent to that context.

The practitioner and the interviewer identify a "focus question" that addresses the problem, issues, or knowledge domain that they wish to Concept Map. Examples would be: "How do thunderstorms form?" or "What is cognitive engineering?"

It is often valuable to begin with an exercise that focuses on content that is very familiar to the person whose knowledge is being elicited. It can be useful or helpful to make a first Concept Map that presents the "big picture," or alternatively, to make a first Concept Map about some very limited subdomain of knowledge. Doing this may take the process away from the topics that are important to the immediate research project goals, but the exercise can be critical in acclimating the participant to Concept Mapping.

The focus question is typically expressed as an unattached node or header toward the upper left corner of the Concept Map space. The explicit presence of the focus question helps to keep the discussion oriented on the knowledge that is most relevant to the problem or question. Stating an explicit question can be very helpful in identifying the most important concepts to include at the higher levels of a Concept Map. In turn, identifying the most important concepts to include at the top of a Concept Map often leads to a refined focus question.

Guided by the focus question, the participant is asked to identify five to ten of the broadest, most overarching, more general, or most important concepts that are involved in the topic. The initial set can be created through a deliberative process on

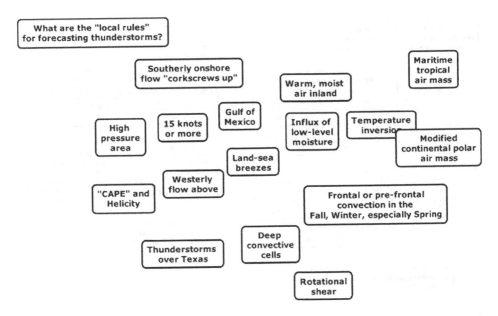

Figure 4.5
A step 1 Concept Map about thunderstorms.

the part of the practitioner or through a free-association process. Thus, a train of thought about thunderstorms might be, "They require low-level moisture," "They can cause tornados," and so on.

Usually, when more than one word is used to label a concept, the participant can consider whether some of the words in the label are also labels for concepts and should be indicated as separate concepts. Our experience in knowledge elicitation is that once concepts are pulled apart, they are usually useful later on to tie other things together and express additional meanings. One should avoid sentences in the concept boxes since this usually indicates that an entire subsection of a Concept Map could be constructed just from the statement in the node.

Figure 4.5 is a step 1 Concept Map. The focus question was selected because the forecasting of thunderstorms was the participant's specialty, and the question focused on the researcher's interest in getting at the local "rules of thumb" used in forecasting. Note that this Concept Map includes nodes that have embedded concepts that would eventually be pulled apart.

Step 2: Set up the "Parking Lot" and Arrange the Concepts   Next, the concepts are arranged in what is called a "parking lot." The concepts are moved around in the

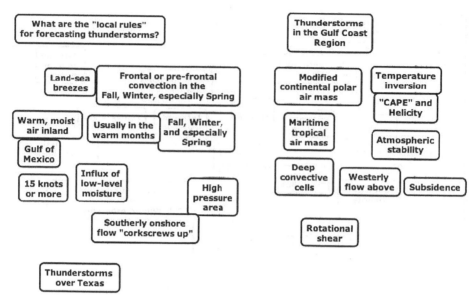

Figure 4.6
The thunderstorms Concept Map at step 2: arranging the parking lot and adding more concepts.

space to place the most inclusive or most general concepts (those that seem to be most important or most closely related to the topic) toward the top of the Concept Map. In addition, more important concepts are added. An example appears in figure 4.6.

Some individuals who are new to Concept Mapping, and some individuals who have had experience at it, have a style that short-circuits step 2. They prefer to lay out a few high-level concepts and immediately begin linking up the concepts (see step 3). This is perfectly acceptable as a style, although it has been noticed that individuals who skip step 2 and jump right to Concept Map construction tend to "dig down into the weeds" prematurely and can lose sight of the larger picture.

Step 3: Begin to Link the Concepts   At this step, the mapper begins to link the concepts. A linking word or short phrase should define the relationship between the two concepts, so that the node-link-node triple reads as a proposition. This step in the Concept Map process, and the importance of the link labels, are both illustrated in figure 4.7. Even for such a simple Concept Map, without the relations, the meaning and communication value disappear. One can guess at some of the relations (in this case, the nodes under "Land-sea breezes"), but even in this simple Concept Map one can

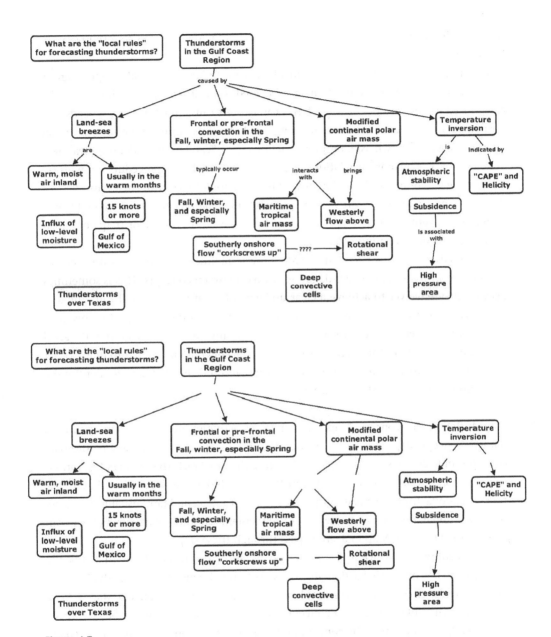

Figure 4.7

Step 3: linking concepts (top panel) and a demonstration of why the linking phrases are important (bottom panel).

see the value of the linking relations (e.g., how is "CAPE" related to "Temperature inversion"?).

It is necessary to try to be precise in identifying linking words, although there is no limit to the sorts of terms that can be used as relational links. They could be categorized in any number of ways. Links can express causal relations (e.g., "leads to," "produces"), classificational relations (e.g., "includes," "is an example of"), nominal relations (e.g., "is known as"), property relations (e.g., "can be," "has defining feature," "consists of"), explanatory relations (e.g., "is a reason for"), procedure or method relations (e.g., "results in," "is done by," "is a way to do"), contingencies and dependencies (e.g., "requires," "often is"), probabilistic relations (e.g., "is more likely," "rarely is"), event relations (e.g., "comes before"), and uncertainty or frequency relations (e.g., "is more common than").

Individuals who are new to Concept Mapping sometimes comment that it is difficult to come up with appropriate words to use as links between concepts. This is sometimes because they have yet to achieve a clear understanding of propositional thinking and relationships between concepts, and it is the linking words that specify this relationship. Once people begin to settle on good linking words and also identify good cross-links (see step 5), they can see that every concept could be related to many other concepts. This can also produce some frustration, and they must choose to identify the most prominent and most useful links.

Step 4: Refine the Concept Map   A number of activities are involved in creating a refined Concept Map. This includes adding, subtracting, and changing superordinate concepts and adding, subtracting, and changing the link labels that express the various subsumption and differentiation relationships. It includes checking to see that all the node-link-node triples express propositions. Good Concept Maps usually are those that have undergone one or two waves of refinement, although with practice one can make a good Concept Map in a single pass through these steps.

The meaning of a concept is represented by all of the propositions that link the concept in a given knowledge domain. Thus, to define the meaning explicitly, it is generally preferred to use a given concept label only once in a given Concept Map. A Concept Map that contains the same concept two or more times can usually be rearranged (this takes practice) so that the concept only appears once. Sometimes this may require reconstruction of other sections of the Concept Map, and usually this leads to general improvement.

As a rule of thumb, if there are more than four or five concepts linked under a given concept, this means that there are latent concepts, or some sort of intermediate level that the practitioner has not yet expressed. Specifying these intermediate concepts often leads to insights, and once the latent concepts are made explicit, they often serve a useful role making it possible to link to other concepts, expressing relations that would have otherwise been difficult or even impossible to express as long as those intermediates had been left tacit. For instance, one Concept Map on weather forecasting included a number of different data types linked as sources of information that support storm forecasting—radar, observation charts, buoy data, satellite images, and so on. In the Concept Map, these data types splayed out under the "Forecasting Products" node like a large fan, consuming a considerable amount of space in the Concept Map. It was possible to split these off according to an intermediate level of conceptualization—data types that were graphic products (radar, satellite images, etc.), data types that involved alphanumeric data (charts, buoy reports, etc.), and data types that came from observations (i.e., sky watching, surface observations). The result was a much "tighter" Concept Map, and the intermediate concepts could subsequently be used to advantage.

Figure 4.8 shows the thunderstorms Concept Map after it has undergone two waves of refinement beyond the version shown in figure 4.6. Note that this Concept Map still needs work—there is a "fan" coming off the "Land-sea breezes" node. There is a node that needs unpacking ("Release of latent heat from moist [saturated] air"). There is a string that needs fixing: "Southerly onshore flow rises and corkscrews up when it meets westerly flow above and induces rotational shear." The node-link-node triple, "Westerly flow above and induces rotational shear" is not a well-formed proposition. Finally, there are some concepts left over from the parking lot.

Step 5: Look for New Relations and Cross-Links, and Further Refine the Concept Map   At this step, the mapper looks for "cross-links" between concepts in different sections of the Concept Map. A good example appears in figure 4.8, where the mapper has created a cross-link between "Land-sea breezes" and "Severe thunderstorms" and has indicated that "Southerly onshore flow" is not the same as "Land-sea breeze." The mapper is considering how to phrase a link between the concepts of "Westerly flow above" and "Subsidence."

This process of refining a Concept Map can go on as one brings to bear additional knowledge. In our experience at Concept-Mapping knowledge elicitation with domain practitioners, it almost always happens that at some point in the procedure, the practitioner says something like, "You know, I've never really thought out this (concept, relation) in quite this way, but now that it comes up. . . ." Here we see that Concept Mapping often serves to elicit knowledge that might have otherwise remained tacit.

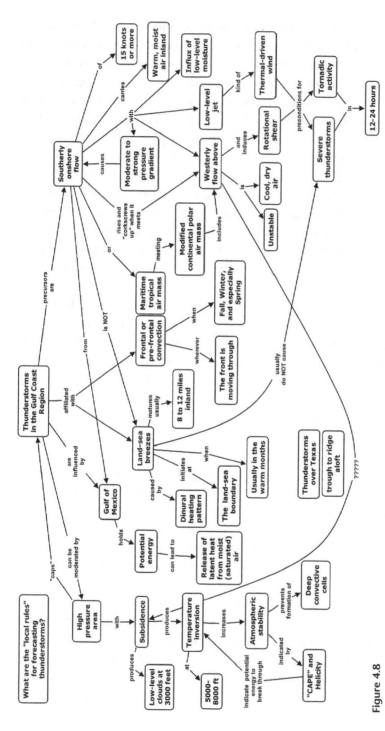

Figure 4.8

Considerable refinement has been made to the thunderstorms Concept Map.

It is important to recognize that a Concept Map is never finished. There is no one right way to make a Concept Map for a given domain or subdomain of knowledge.

Research suggests that a Concept Map that has been created and refined by one expert can expect to have about ten percent of its propositions altered when the Concept Map is evaluated by some other expert (Hoffman, Coffey, and Ford 2000). This is not because experts disagree (although they can). Rather, it is wordsmithing—a reflection of their differing emphases, their judgments of what is important, and the subtleties of word choice (e.g., "promotes" versus "causes"). As one's understanding of relationships between concepts changes, so will the Concept Maps. Conversely, the process of Concept Mapping almost always leads to new understandings and insights on the part of the domain expert who is building the Concept Map.

Figure 4.9 shows the thunderstorms Concept Map after a further wave of refinement, which involved a slight change to the focus question.

Step 6: Build the Knowledge Model A set of Concept Maps all on a particular topic and hyperlinked together is referred to as a knowledge model. Resources are another important feature of knowledge models, and the process of adding resources to Concept Maps should be considered integral to Concept Maps as a form of CTA.

Note in figure 4.4 that there are small icons appended immediately beneath some of the concept nodes. These all represent hyperlinked resources. Resources can be many forms of digital media, including text, detailed examples, images, charts, links to PowerPoint presentations, Web pages, digital video, and the like. Resources can be URLs that go out and grab data fields and return them for presentation within the context of the Concept Map. Resources can be text pieces that go into detail about the concepts to which they are appended. They can link to operational manuals, standard operating procedures documents, or forms that the practitioner needs to complete. They can present case studies that illustrate and concretize the concepts. In a knowledge model on weather forecasting (Hoffman, Coffey, and Ford 2000), resources include URLs that can take the user to the real-time data that are involved in the practitioner's task (e.g., radar, satellite images, etc.). Data are presented in the context of Concept Maps that provide the explanatory glue that makes the forecasting process hang together.

Our experience has been that domain practitioners almost always keep a file of "special" resources. One weather forecaster, for instance, kept a file of hard-copy radar and satellite images from previously encountered difficult forecasting cases. Such material is a gold mine for the knowledge model because it will contain resources of great potential value to learners.

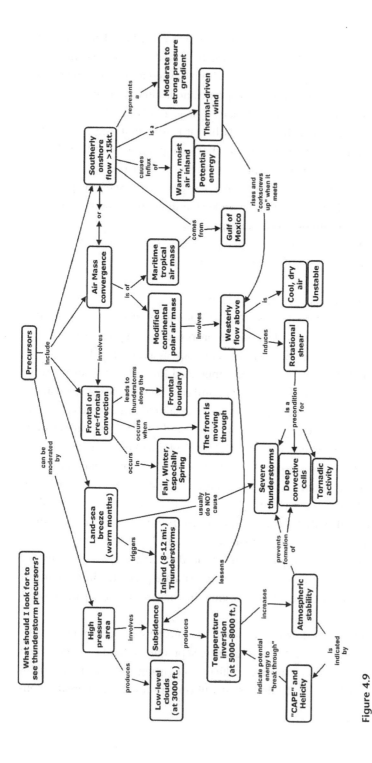

Figure 4.9

The refocused and refined thunderstorms Concept Map.

In addition to having resources added, Concept Maps on a given theme are often hyperlinked together. So, for instance, a Concept Map about "Gulf of Mexico Effects" on regional weather might have concept nodes referring to "fog," "thunderstorms," and "hurricanes," each of which is the topic for its own Concept Map. Through the hyperlinks, the user can navigate the knowledge model. In addition, the top node in each Concept Map in a knowledge model has a hyperlink to a "map of maps." Through this, one can get from anywhere in the knowledge model to anywhere else, mitigating the problem of "getting lost in hyperspace."

The general procedure for eliciting and representing practitioner knowledge presented here carries over to knowledge elicitation with teams or groups.

Team or Group Concept Mapping

As with all teamwork, optimal size of the team is never easy to preordain. Experience has shown that the optimal size for Concept-Mapping knowledge elicitation groups is usually five or fewer. With groups of more than five, especially if one of the participants seems controlling or seems intent on running an agenda, disagreements can occur, requiring patience, finesse, and cat-herding. Furthermore, individual differences in style, personality, and preexisting interpersonal relations can swamp any generalizations about group size. The group needs to be large enough so that most of the important subdomain knowledge and/or experience is represented in the group, but not so large as to make whole-group discussions difficult.

There are a number of different ways in which Concept Mapping can be conducted with groups of participants. Teams can discuss the Concept Maps made by individuals. A team might assemble a "Global Concept Map" for their domain or organization. A team leader might define the key question and create a preliminary global Concept Map, and the team can work off of that. Alternatively, a team might be divided into subteams to develop Concept Maps for subdomains.

What Makes for a Good Concept Map in the CTA Context?

The value of Concept Mapping as a knowledge elicitation procedure lies in achieving a clear, precise description of domain knowledge. Good Concept Maps do that well.

A good Concept Map is comprehensive relative to its focus question and top node. It is important to ensure that all of the concepts associated with the topic and pertinent to the CTA goals are included in the Concept Map. Sometimes basic and important concepts are overlooked. But, this raises the question of how big a Concept Map can be. One heuristic is that there is a threshold, that Concept Maps need to be large or complex enough to maximize the chances for identifying significant cross-links, and yet not so large as to suggest that they be split up into submaps. Some Concept Maps are quite large, consuming a great deal of space, but these require lots of vertical and horizontal scrolling. In this regard, the heuristic is to rely on the human scale of the typical computer monitor. A rich Concept Map, one containing on the order of forty concepts and forty-five or so propositions can, with skill and some finesse, be comfortably fit into the screen space.

A good Concept Map manifests global relevance. Since there is a significant element of associational thinking in the creation of a Concept Map ("Concept 1 makes me think of Concept 2"), it can happen that the mapper introduces concepts that are of relatively low relevance to the topic at hand. Judgments must be made regarding the relevance of every concept to a particular topic. The Concept Map in figure 4.8 includes a node ("Thunderstorms over Texas") left over from the parking lot that may be thought of as being a bit too far off the central topic. Likewise, the concepts in figure 4.8 under "Gulf of Mexico" were deemed unnecessary for the refocused and refined Concept Map (figure 4.9). That material was eventually split off into a separate Concept Map just about "Gulf of Mexico Effects."

A good Concept Map also has the right "granularity." One type of granularity problem is when a Concept Map dealing with very specific concepts has a few unnecessary and very broad concepts—in places the Concept Map seems to go off on a tangent. Conversely, granularity also becomes an issue when a Concept Map on a broad topic has some overly specific or detailed concepts—in places the Concept Map goes too far down into the weeds relative to its top node or its focus question.

There are always trade-off decisions that have to be made concerning comprehensiveness, relevance, and granularity. In an effort to ensure that all the important information is included (to be comprehensive) it is possible that ideas of minimal relevance might be introduced (i.e., in places the Concept Map might go too far down into the weeds or might go off on tangents). Likewise, in an effort to ensure that only relevant concepts and their relations are included, some concepts that are important might be missed.

Summary

CTA involves capturing what practitioners know about their domain: its concepts, principles, and events. We can think of no CTA process or project in which the CTA researchers did not have to elicit and then represent at least some domain knowledge. This chapter reviews the procedures and applications of Concept Mapping as a proven knowledge elicitation method for the efficient elicitation of practitioner knowledge.

Concept Maps involve labeled nodes and links, but Concept Maps differ in important ways from other types of diagrams that utilize combinations of graphical and textual elements to represent or express meanings. Concept Mapping supports the practitioner's effort to reach for crystal clarity about what he or she wishes to express. In Concept-Mapping knowledge elicitation, the researchers help the domain practitioner build up a representation of domain knowledge, in effect merging the activity of knowledge elicitation and the activity of knowledge representation.

Needs?

all maps turn into databases?

5 Incident-Based CTA: Helping Practitioners "Tell Stories"

One of the most powerful knowledge elicitation methods available to Cognitive Task Analysis (CTA) practitioners is to probe actual incidents. People tell us about all kinds of details, challenges, subtle cues, background influences, and strategies that might never come to light in a general interview or a controlled simulation. Skilled decision makers have had many different experiences; that's how they formed their knowledge and honed their skills. Their stories can be a doorway into that experience.

Take weather forecasting as an example. Experienced forecasters have always traded insights using case studies of particular weather events. Case studies appear regularly in journals such as *The Monthly Weather Review* and *Weather & Forecasting*. Senior forecasters love to tell stories about their first successful tornado forecast or the first time a fog forecast "busted." Some of the stories[1] they remember are rich in detail:

It was midwatch. Before the Mobile [Alabama] radar painted it, we knew only that there was a southwest air flow and clouds down over the Gulf south of New Orleans. It was a bad system. It was southwest, about 100 miles south of the mouth of the Mississippi River. It was a big storm cell. It was moving north-northeast. I knew it would hit close by and would affect our area. After one hour I knew it would qualify as a supercell. When it crossed the mouth of the Mississippi, the Weather Channel said, "Look at this supercell!" We'd been looking at it for over an hour. Slidell and Mobile radars were getting good reads on it. I kept extrapolating the track via the NEXRAD radar. I watched it loop by loop. Bad cells tend to turn to the right but they can sometimes turn to the left. If it is upstream of you, you are not going to take your eyes off of it. I knew it would hit at about 3:00 AM. When it got 40–50 miles south-southwest, I realized it would track 20 miles east, right about at Smith Field. I called Smith Field at about 1:00 or 2:00 AM. They had a young forecaster there, just out of [his first duty assignment], and hadn't worked any severe weather. I asked if he was aware of the supercell heading toward him. He said, ".... What?!"

Many forecasters can even pull out their "special" files of the records they have kept of interesting and tough cases they experienced. They thrive on the details and have clear awareness of each of the lessons they learned. Once a forecaster gets going on a favorite story, he or she can take an hour or more simply to lay out all the details.

With luck, apprentice forecasters may get a chance to learn from the experiences of the senior practitioners. Here we see, of course, one of the applications for CTA—knowledge sharing and training. The lessons learned that are contained in stories also suggest leverage points, perhaps for new decision aids. Managers can also use stories to appreciate what makes their staff members expert and to take that into account in running an organization.

What sorts of things can the CTA researcher find in stories?

• The cues and patterns that experts perceive:

I could see the air pressure falling and knew I could put out a warning for strong winds. [The lightning network] showed a ring of lightning around the "Low" pressure center. This was unusually symmetrical, but showed that the Low was well organized. From a hand plot of buoy data about air pressure, I could plot the front, the Low's position, movement, rate of movement. I did about one plot per hour, about six or eight in all. Enough to know that the warning had to go out and then two or three more plots to show that it really was out there.

• The rules of thumb they have devised:

It was a warm air mass over a cold air mass condition, which trapped the fog. Gulf of Mexico moisture was coming up due to high pressure over the Gulf. The airport was just high enough in elevation to condense the moisture and form fog. The forecasting problem was if and when the fog ceiling would raise enough for flights. [Trainee pilots] needed to fly. And it was midweek so they were busy. We'd look over the [airfield] toward the downtown hotels and use the hotels as ceiling indicators. The downtown is 15 miles away. If you could not see the top of a certain building, you knew the ceiling was 800 feet. We knew from the visibility of the hotel floors what the ceiling was, and when it got up to 800 feet. There were other rules of thumb. If you could not see the airport tower you knew the visibility was less than 3/4 of a mile. You use what you can. The pilots kept bugging me so I had to keep monitoring the situation—satellite loop, visibility, every 5 to 10 minutes, observations out of the area airfields. By 1:00 PM I knew no one would fly.

• The kinds of decisions they have to make:

I came on midwatch duty Saturday evening. The National Hurricane Center (NHC) had Hurricane Georges tracking west-northwest. The computer forecasting models had it going every which way after landfall. The NHC had the wrong track. They were wrong on where the eye of the hurricane was. We could see it on radar. You could see the eye wobble on the satellite image loop and the radar loop. The eye was running in and out and sometimes was defined and sometimes was not. We looked at buoy data every few hours and did our own charts. [The participant in this interview had kept the originals and pulled them out of a file drawer.] The NHC shifted the hurricane track a little to the east out to Gulfport Mississippi, but we were leery about that track. They were still off. The NHC had it shifting northwest to Louisiana, more of a westward track. But we could see it heading due north toward Biloxi. We had to go with the official forecast.

• The features that make decisions tough:

The analysis of the upper atmosphere showed an area of turning winds. My goal was to try to figure out what would happen. If it kept moving through, nothing would happen, but if it didn't, you get caught with your pants down. The wind shift implied that something was happening. A novice would have missed it. The region of maximum wind curvature was at the top of the high pressure ridge so it would not show up so much. The turning of the winds was enough to me. Then you look for support and weigh all the factors. [The cirrus] was not enhancing, but it also was NOT going away. There was moisture at the upper levels. Water vapor imagery showed no slot of dry air associated with the front, implying there was no instability. But I was seeing that there was potential here. There must have been something balancing the wind curvature. The upper-level wind should have changed as it was going down the ridge.

• The features that make cases typical:

It was March. Before arrival [at the weather station] I was skywatching. It was not a blue or gold sunset. I saw cirrus to the southwest, anvil cirrus blowing off the tops. You can see this even though the main clouds might be 100–200 miles away. This confirmed that there was energy out there. There were not enough data yet. We had to query the buoys. This was a textbook case. A stalled front off the Texas coast. You look out to the southwest and if you see any approaching trough, vorticity, or a vorticity maximum, any Low or wave on the front will develop one or two storm systems. It is taught in the School and is discussed in the Local Handbook. But you still need to experience it firsthand a few times. If you get burned once, then you learn.

• The features of rare cases:

Maintained gale-force winds require major storm systems. This is rare. The storm of March 1993 hit western Florida with 112 mph winds. That situation was similar to this one—everything lines up perfectly. But major storms out of this scenario are rare. These were minor storms. I was asking myself, were the Lows intensifying and moving eastward? Intensification would imply a need to upgrade the warning. Would people need to do preparations at the [airfield]? It was not a routine situation since it is not usual to get a heavyweight supercell at midshift. This was a standard scenario in terms of the storm development and dynamics, but not standard in terms of the time of year and time of day of the storm. Fast-moving cold fronts coming from the west usually determine our winter weather—storms and small lines of storms. A big cell developing in a southwest flow is rare for winter in the Gulf region.

These kinds of information are contained in stories that can be elicited in any domain. Military leaders, project managers, nurses, sales personnel, firefighters, even consumers can describe incidents for the CTA researcher to study. We have developed the Critical Decision Method (CDM) to learn from specific incidents (Hoffman, Crandall, and Shadbolt 1998; Klein, Calderwood, and MacGregor 1989). Many CTA researchers use the CDM for conducting incident-based interviews (Blandford and Wong 2004; Ebright et al. 2003; Klein and Armstrong 2004; Militello and Crandall 1999; Omodei, Wearing, and McLennan 1998; Readinger, Ross, and Crandall 2004; Thordsen 1991;

Wong, Sallis, and O'Hare 1997). This chapter describes how to conduct an effective CDM interview.

How can you go about "grabbing" the power that resides in the practitioner's experience? One way is to ask the practitioner how they do what they do, in a general or abstract way, as in "How do you predict tornados?" or "What's involved in doing the forecasting job?" This type of general question serves to divorce the practitioner's knowledge and skills from their lived experience.

In contrast, the CDM deliberately avoids generic questions of the kind, "Tell me everything you know about X," or "Can you describe your typical procedure?" Such generic questions haven't been very informative. One reason is that complex domains usually don't have simple, general, or typical procedures. Even if the work seems to be typical, we usually find many alternative types of action sequences even for routine tasks and situations. Furthermore, procedures change depending on style, the status of the equipment, and the skill level of the practitioner.

We came up with the idea of conducting the knowledge elicitation by asking people to tell us about previous incidents as a practical solution to a data collection problem. In a study of firefighters (Klein, Calderwood, and Clinton-Cirocco 1986), we had the notion of "shadowing" the firefighters, riding with them to fires, and interviewing them at the scene of the fire as the incident unfolded. We wanted to stand side by side with the firefighters and get them to "think aloud" (perhaps prompted by a few questions). But what seemed like a great way to get field data turned out to have a major glitch: Firefighting is an "on call" occupation, and there were stretches when there weren't that many calls. We soon realized we were likely to spend a lot of time (and money) waiting around for fires to happen so we could collect data, and that we weren't going to get very far by simply relying on observations. Instead, we used the downtime to collect firefighters' stories about some of their past experiences, using an adaptation of Flanagan's (1954) Critical Incident Technique. The retrospective method—asking people to tell us about previous incidents—arose out of necessity.

These CDM interviews rely on retrospection. In conducting CDM interviews we have to face the possibility of memory loss and distortion when significant time has passed since a to-be-recalled incident occurred. That is one of several reasons for probing non-routine, challenging events. By their nature, challenging events are going to call for whatever expertise a person can bring to bear on the situation. They evoke focused attention and depend on full use of skills. The outcomes often have more riding on them. For all these reasons, they are more vividly recalled than routine events.

This chapter covers two major topics: First, we describe the steps of the CDM—a method for mining people's real, lived experience and getting inside their heads to understand incidents from their perspective. As we walk through the CDM interview process, we take you inside the interview to let you see an experienced CDM interviewer at work. Second, we discuss boundary conditions—when incident-based methods are most useful and when they are less so. We also describe variations of the CDM and how to adapt the method to different settings.

The Critical Decision Method (CDM) Procedure

The CDM is an intensive interview that often takes as long as two hours. In some cases such as weather forecasting where incident memories can be very rich, the CDM steps can even be broken up and conducted over several sessions. The CDM interview is conducted by two researchers. One interviewer acts as the primary facilitator, but also takes notes. The second interviewer is primarily responsible for taking a good set of notes and keeping track of the overall plan for the interview.

After making introductions, gathering some demographic information, and spending a few minutes establishing rapport, the main portion of the interview is carried out by making several "sweeps" through an incident. Each sweep constitutes a pass through the incident and builds on the previous sweep(s). Each is focused on eliciting specific types of information. At the end of the interview, the research team has a thorough understanding of the incident from the perspective of the interviewee.

In a CDM interview, the researcher tries to elicit information about cognitive functions such as decision making and planning and sensemaking within a specific challenging incident. The overall data collection strategy is to gradually deepen on critical cognitive points by making multiple passes through the incident. The research team has to get the story of the specific event and understand the cognitive demands of the task and setting.

The interview is conducted in four phases, or sweeps: (1) Incident Identification, (2) Timeline Verification, (3) Deepening, and (4) "What If" Queries (see figure 5.1). Each sweep uses different kinds of probes and perspectives and helps the participant recall events in greater detail.

In the following sections we provide a description of each sweep followed by a description of that portion of the interview from the researcher's perspective.

Sweep 1: Selecting an Incident

The initial CDM step is focused on identifying candidate incidents and selecting an appropriate incident for deepening. The precise type of incident will depend on the

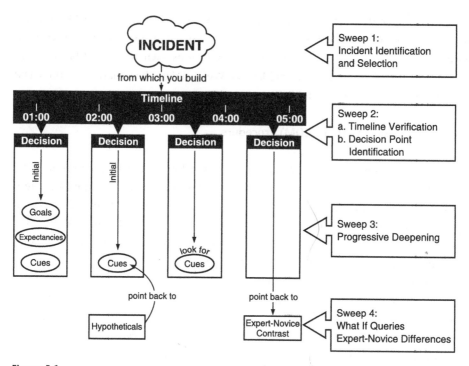

Figure 5.1
The structure of the Critical Decision Method (CDM) procedure.

nature of the project and goals for data collection. Traditionally, CDM has been used to examine nonroutine, challenging events, because these tough cases have the greatest potential for uncovering elements of expertise and related cognitive phenomena. For example, the interviewer might ask the participant to recall a time when his or her skills were particularly challenged, or when knowledge and experience really made a difference in the way the incident turned out. Generally, in this sweep you want to identify an incident that will contain cognitive components that go beyond background and routine procedural knowledge of the domain and that will enable you to learn about those components that characterize skilled performance and expertise.

Once you identify a good candidate incident, ask the person you are interviewing to provide a brief account of the story, from beginning to end. The initial account, and the content of the story, is the foundation for the rest of the interview. Table 5.1 offers a look at sweep 1 from the interviewer's perspective.

Table 5.1

Interviewer's perspective: What you're doing at this point in the interview—Sweep 1

Sweep 1: Incident Selection

Identifying a good incident has a number of elements:

- You are listening for a couple of key indicators of relevance. First, did the person you are interviewing have a role as a "doer/decision maker" in this event? Given the particular type of skill you are interested in, was this person in that role? For example, for a study of fireground commanders, the seasoned firefighter who was first on the scene and held command until backup arrived may have an interesting incident to tell. The firefighter who *witnessed* an incident involving exceptional command skill can't help you if he or she was not in the command role.

- What about telling participants ahead of the interview session that you are going to ask them for incidents? On the face of it, this seems like a way to save time for everyone. But it is risky. When they have advance notice, people mull over incidents. They are likely to rehearse a bit and in doing so they may begin to alter the story. They reorder events so that they will "make more sense." They discard the parts that seem not to "fit" or seem irrelevant. They smooth out all the edges, and leave out the embarrassing part where they made a mistake. These are exactly the details we want and need—so it's better to surprise them and get the story fresh.

- In asking for an incident, if you stress the unusual you are likely to get exactly that: weird stuff. Or you may get "critical decisions" but not of the type you are interested in. For example, a critical parameter for firefighters is whether or not life was lost. In the original firefighter studies, stories about those incidents were dramatic, often tragic, but they did not necessarily produce the kind of decision elements we were looking for. The issue is whether the person's decision making (or other cognitive event) had a direct impact on the outcome. If it did not, then the incident is probably not a good one for our purposes. (It is for this reason that the CDM was given its name.)

- When you are working in a new domain, you may find yourself wrestling a bit with whether an incident is worthwhile or not. You may need a few interviews before you have a better feel for what sorts of incidents the term "critical" is likely to elicit. There may be false starts and a need to use alternative opening queries.

- Be willing to sit quietly and let the person you are interviewing think about your question, even struggle a bit to come up with an incident. Do not rush them. If they say, "I can't think of anything," you might reply, "Let me say again what we are after." Repeat what you said before and add a bit of description, or rephrase your opening query in a slightly different way, and again give them time to produce something for you. Sitting in silence can be very hard, but the ability to tolerate silence is a key interviewing skill.

- Whatever your criteria, most people are going to have only a handful of incidents that fit it well. They won't need to sift endlessly through their whole past. In choosing an event to talk about, they will say, "Well, there was this one time..." They will give you an overview, an outline. If it doesn't sound worthwhile, you might say, "That sounds interesting, but we are looking more for incidents that..." and restate your criteria with some rephrasing. "Could you think of one that has more of that flavor to it?" Don't screen all possible entries. Once you hit one that sounds good, go with it. If it is the first one the person brings up, that's fine.

- What the person tells you gives you the content of the story. How they tell you the incident gives you the "bones," the basic structure, for the entire interview. In addition to the content, they have given you a sequence, organized into a series of segments. These incident accounts come to you initially as spoken stories that have an inherent structure and rhythm. Rhythm is about pauses—where there is silence in relation to where there is sound. Listen for the pauses, for where the person's voice falls for a moment before the next piece. Listen for the turning points, when the action or the entire scene changes. Listen for the words: "So then,..." These

Table 5.1

(continued)

are meaningful demarcations in the event. The sequence, the segments, and the pauses give you the frame of the story, dividing it into meaningful parts.

- You may need to move the person along in order to get through the initial account. People are usually eager to help, but they don't know exactly what sort of information you want, or at what level of detail. As they get into telling the story, they may dive down into the weeds, or they may wander off on a tangent and begin instructing you about standard operating procedures and general principles. You can help keep them on track by saying, "We're very interested in that, and I'd like to talk more about it in a bit; for now, can you give me a quick overview of this particular incident, so I have a sense of what happened from beginning to end?"

- In providing their view of the incident, the person defines the beginning and the end of the story. It can be informative to prove the beginning and end they provide. We often wonder, and sometimes ask, "What was happening right before this?" Or, "How did this turn out eventually?" Sometimes, what happened just prior to the person's starting point contains critical information for understanding the event itself. Sometimes the story has a second ending that provides a whole new perspective on the incident and the participant's role in it.

Sweep 2: Constructing a Timeline

The second sweep is aimed at getting a clear, refined, and verified overview of the incident structure, identifying key events and segments. This is a key step, because that structure will provide a crucial framework for the remainder of the interview. In addition, the person being interviewed often begins to recall events in greater detail and more fully relives the event. If you were observing a CDM interview, you might have a hard time telling where sweep 1 ended and sweep 2 began. Once you have identified an incident that appears to fit your project goals, and you have the initial incident account, it is appropriate to start verifying the timeline.

During sweep 2 the interviewer works with the participant to expand the initial, brief account of the incident. As the interviewer diagrams the sequence of events the participant might notice that something is out of sequence or that an event is missing and offer corrections and additional details.

Figure 5.2 contains an example of a timeline developed during an interview with a fireground commander. Clearly, the incident depicted here was challenging, and events were developing very quickly over a short time (approximately twenty-five minutes). Notice that the timeline is not laid out in equal intervals. The time hacks reflect timing of events as they actually occurred rather than fitting the incident into preset, regular time units.

In diagramming the timeline, the critical points (sometimes called "decision points") are when the practitioner experienced a major shift in his or her understanding of the situation or took some action that affected the events. They are the critical junctures in

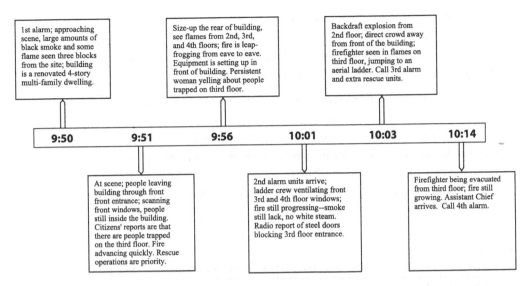

Figure 5.2
Example of a timeline from an interview with a fireground commander.

the event, turning points where the situation could have been understood or acted upon in several different ways (and not just the times when "good" decisions were made by the participant). Table 5.2 provides a look at sweep 2 from the interviewer's perspective.

As the sequence and duration of events, actions, perceptions, thoughts, and decisions emerge, the interviewers and participant arrive at a shared view of the facts of the case from the participant's perspective. Working through the incident in this way, interviewers are able to clear up inconsistencies, identify gaps, and fill in missing elements. Skilled interviewers are also able to begin tagging key segments and decision points to probe later on. As participants go back through their account, additional details emerge.

After interviewers have elicited and documented the incident and clarified and verified it with the participant, it is time to move to the next phase of the CDM interview: deepening.

Sweep 3: Deepening

This sweep is the most challenging, but also the most fun part of the interview. This is where you have the opportunity to get inside the expert's head and look at the world through his or her eyes. From the interviewer's perspective, the guiding question is:

What is the story behind this story? Based on the first two sweeps, I know what happened, who did what, and I know a bit about their role in the event. But what did they know, when did they know it, how did they know, and what did they do with what they knew? That's what Sweep 3 is designed to figure out.

Going beyond the time elements and basic facts of the incident, what were the participant's perceptions, expectations, goals, judgments, confusions, and uncertainties about the incident as it unfolded? What was he or she concerned about? What other options did he or she consider in making decisions? What information did he or she need and how did he or she get it? Critical Decision Method probes are used to deepen the understanding of the event—to build a comprehensive, detailed, and contextualized account of the incident from the decision maker's point of view.

Table 5.2

Interviewer's perspective: What you're doing at this point in the interview—Sweep 2

Sweep 2: Place the Incident on a Timeline

▪ A good way to start building a timeline is to go over the initial incident account, saying it back to the participant exactly as they have told it to you. Hold a mirror up for them. Ask, "Do I have the sequence and the details right so far?" Let them hear how the story sounded to you. They may realize something is out of sequence, and offer correction, and some additional details. Why it matters: the purpose here is for the two of you to agree on the overall incident. It's also the point at which the participant's memory begins to really engage. Inserting your language and your version of the story will muddy the waters. You want a clean, clear version of their incident as the basis for the interview, rather than your version of their version. Mirroring their account also shows that you are paying attention, and builds trust and rapport.

▪ What sort of timeline should you elicit? Do you need detailed and specific timing information? An approximate timeline? Are you interested in time (duration) or in timing (synchronicity, sequence of occurrence)? Perhaps it is enough to identify the sequence of events, because specific time designations do not have much meaning in a particular situation. In interview data elicited from NICU nurses (Crandall and Getchell-Reiter 1993), some incidents lasted less than three minutes, while others lasted several weeks. Interviewers had to decide in the moment what a useful time scale was going to be.

▪ Once you've settled on the right structure, the task is to overlay segments of the incident and key decision points on the timeline (or event line, or map). Typically we do this by creating a representation on a whiteboard or on 11 × 17 paper so there is plenty of room to write and work. You want the participant to be able to see what you are doing, what segments you are marking, and how you are labeling them. Engage him or her in the task by asking, "Do I have this right? About where on the timeline should we put _____?" The point here is not to make something that looks good (you can clean it up later if you like) but to generate an organizing framework that will help to keep you on track for the rest of the interview.

▪ You may have some pretty solid ideas about where the key decisions and situational shifts occurred. It is a good idea to get input from the interviewee at this point. You may think, "This is a key decision," but he or she may say, "A guy right out of training would have known to do it that way, or he wasn't paying attention. That's standard operating procedure for this kind of situation." The decision clearly was important to the outcome, but it is not a critical decision in the sense that there was any other choice to be made.

Table 5.3
CDM "deepening" probe questions

Cues	What were you seeing, hearing, smelling, noticing etc.?
Information	What information did you use in making this decision or judgment? How and where did you get this information, and from whom? What did you do with the information?
Analogs	Were you reminded of any previous experience? What about that previous experience seemed relevant for this case?
Standard operating procedures	Does this case fit a standard or typical scenario? Is it a type of event you were trained to deal with?
Goals and priorities	What were your specific goals and objectives at the time? What was most important to accomplish at this point in the incident?
Options	What other courses of action were considered or were available to you? How was this option chosen or others rejected? Was there a rule that you were following in choosing this option?
Experience	What specific training or experience was necessary or helpful in making this decision?
Assessment	Suppose you were asked to describe the situation to someone else at this point. How would you summarize the situation?
Mental models	Did you imagine the possible consequences of this action? Did you create some sort of picture in your head? Did you imagine the events and how they would unfold?
Decision making	What let you know that this was the right thing to do at this point in the incident? How much time pressure was involved in making this decision? How long did it take to actually make this decision?
Guidance	Did you seek any guidance at this point in the incident? How did you know to trust the guidance you got?

Using the timeline and working from notes of the interview so far, the interviewer takes the participant back to the beginning of the incident and moves through it once again, taking the story one segment at a time. For each segment, the interviewer probes for additional detail and the participant is encouraged to elaborate on and deepen the incident account. During this sweep, interview probe questions are aimed at eliciting the cues and information available in the situation, the meaning they held for the participant, and the specific cognitive processes and functions they evoked. Table 5.3 contains the probes we have developed and routinely use for this sweep. Over the years, we have refined and revised the particular questions. Feel free to modify or add to this list as you discover useful questions and probes.

The interviewers will certainly not ask all of these questions for every key segment or critical point that is identified on the timeline, nor will they necessarily ask questions about every topic contained in table 5.3. Knowing what probe to ask, when, and why, is a skill that develops with practice. Generally speaking:

• If a critical point on the timeline involves making an observation, the probes about information and cues are useful.
• If a critical point involves assessing or making sense of a situation or projecting a situation into the future, then probes about assessment and mental models are useful.
• If a critical point on the timeline involves making a decision, questions should be about decisions, obviously, but also about goals and options.
• At points where the story seems to refer to the participant's knowledge, then probes about the basis of choice and about experience are useful.

One of the interviewing skills that comes with practice is figuring out how to get good responses. Interviewers have to be ready to ask the same question in a variety of ways, because a probe that works well for one person may draw a complete blank from the next. When a probe doesn't elicit the information you expect it to, you have to know the reason for asking the question in the first place in order to come at the issue from another direction. Sometimes probing in this portion of the interview is like knocking at a closed door. You can knock once and walk away if the door remains closed; or you can knock again, maybe several times more, to see if you get a response. Table 5.4 offers a look at sweep 3 from the interviewer's perspective.

People sometimes evince physical or emotional reactions that suggest they are very much "in the moment." Firefighters begin to sweat. Pilots jump up, weaving their arms through the air and angling their hands to demonstrate a flight maneuver. People grab paper and pencil and start sketching the scene at a specific point in the incident to show movement of equipment and personnel. Weather forecasters draw simple charts and diagrams showing fronts and other features of weather situations. Sometimes people laugh at themselves, at how excited and involved they have become in telling us about the incident. Sometimes they choke up. Sometimes they cry. In these moments there is a profound sense of the participant reliving the incident and reporting on it as it unfolds. They are more "there," in that other place and time, than they are "here." Sitting in witness of this surging tide of memory can be an extraordinary experience—fascinating, intense, sometimes very moving. It can become difficult, sometimes to keep writing, to stay in one's role as an interviewer and data collector. Sometimes the right thing to do is to put down your pencil and be willing to give your full attention to this person and his or her story, to share the recollection of a moment in time and space that had great meaning for this person. Eventually, the inter-

Table 5.4
Interviewer's perspective: What you're doing at this point in the interview—Sweep 3

Sweep 3: "Deepening" Using Cognitive Probes

▪ Your task at this point is to discover the story behind this story. Based on the interview so far, you know what happened and who did what. But what did they know, when did they know it, how did they know, and what did they do with what they knew? That's what you are there to figure out.

▪ Your specific questions and probes will depend in part on the goals of the project: what the key issues are, what aspects of expertise and/or the situation you identified as important in the preparation and framing activities. How you decide to explore the incident will also depend on what you heard in the initial account. What caught your ear? At what points did the SME say things like *"We just knew..."* or *"My gut told me that..."* or *"It was obvious that..."* (As the interviewer, you may find yourself thinking, "It may be obvious to you, but it's not obvious to me, at least not yet.") The interview guide and the generic CDM probes are there to support active search and discovery, not to constrain or bound it. What's bothering you about the incident? What amazes you that you want to understand? Your own questions and curiosity are a critical compass for directing the interview.

▪ The incident account itself provides guidance and direction. Regardless of how the story has been segmented, take these chunks one at a time and work down through the layers of language and memory until you know everything about that part of the incident that the SME can tell you. Know it in its details and in what exists below its surface. The cognitive elements of this person's experience, of how she or he thought and acted inside the event, are under the surface of the story, waiting to be discovered.

---

viewee will take a deep breath and come back into the present, and you will pick up your pencil, and the interview will move on.

However, it doesn't happen like that every time. Even the most skilled and experienced of us have encountered interviews that yielded little, where participants are unable or unwilling to say what they know and to share their experience. It doesn't mean that the method is wrong, or that your skills are lacking. It just happens that way sometimes. Our suggestion is, if you have three interviews in a row that seem to get stuck, you should examine your interview strategy, your probe questions, and perhaps your choice of methodology (see the following section on boundary conditions).

At other times, participants can fly off in ten different directions, giving detail that is not relevant to the project. It is not unusual for participants to drift away from the specific case and shift into a generic discussion of how things are usually done or give a tutorial about the domain. One of the indicators that this is happening is a shift from first to third person pronouns (e.g., "You can always tell when things go wrong..."). One of your tasks is to keep the participant focused on the particular incident—the facts of the specific case. This doesn't mean you aren't interested in the participant's general knowledge—if it influenced this case, it is meaningful. But if they are launching into a tutorial on basic procedures, then it is time to shift the focus of the discussion back to the specific case. You can easily do this by asking, "Is that what happened in THIS case? Is that what you did THIS time?" It is your task to "steer" the interview, but it's important to do this with finesse and respect.

This sweep of the incident should yield a portrait of the participant's cognitive experience, skills, and knowledge. At the end of sweep 3, you should have a detailed, specific, and fairly complete picture of each segment of the event and of the overall incident.

Sweep 4: "What If" Queries

The final sweep of the CDM interview provides an opportunity to round out the interviewer's insight into the participant's experience, skill, and knowledge. Once again using the incident as a starting point, the interviewer poses various hypotheticals about the incident. These may be asked about the overall incident or about particular segments or aspects of the incident. One possibility is to invite the participant to speculate on how his or her responses in the event might have differed, or how the outcome might have been altered. Some suggestions for probes to use in sweep 4 are presented in table 5.5.

The "What if?" probes illuminate expert-novice differences and potential vulnerabilities for error in the domain. The probes allow you to expand the interview, using what actually happened as a springboard. Another way to expand the interview is by using props. Pictures, objects, photos/drawings, mockups, and storyboards all may be useful for depicting a hypothetical case or as a basis for posing several different hypothetical configurations. Alternatively, props can be useful for representing a concept that you want to get reactions to or ideas about. For example, we used props to enhance our questioning during a project examining women's concepts about osteoporosis. All the

Table 5.5
CDM probes for sweep 4

Expert-novice contrasts	If a novice had been in charge at this particular point in the incident, what type of error might she or he have made and why? Would they have noticed what you noticed? Would they have known to do X?
Hypotheticals	If [key feature] of the situation had been different, what impact would it have had on your decision/assessment/actions/plan?
Experience	What training might have offered an advantage in this situation?
Aids	What knowledge, information, or tools/technologies could have helped?

women in the study took nutritional supplements, including a calcium supplement. At a particular point in the interview we showed them an array of calcium supplements. Then, we showed them a bone (it was actually plastic, but it looked real). We said, "You've told us that calcium helps make your bones stronger. Can you describe for us how the calcium gets from here (the supplements) to here (the bone)?" Using these props was a much more effective way of eliciting their mental models of the role of calcium in bone health than simply posing an abstract question.

One of the key decisions that interviewers must make is how to allocate time for each of the sweeps and whether sweep 4 is essential to project goals. If an incident is very rich, two hours can seem barely adequate, and you may decide to forego most (or all) of sweep 4. There may be one question from sweep 4 that is essential to get to in the interview, and others that are considered extras that may be used if there is time. One of the CTA skills that develops with practice is the ability to think on your feet as the interview progresses, to figure out what information is most important and how to allocate the time you have.

We have found CDM interviews to be surprisingly intimate encounters. People share their experiences and sometimes gain new insights into what happened or realize aspects of their skills and knowledge that they may not have fully appreciated before. They may learn how they actually made critical decisions that they have been thinking about for years. One of the pleasures of doing CDM interviews is witnessing these moments of self-discovery. We have also found that it can be important not to end the interview session too abruptly. Instead, you should leave a few minutes to debrief, to answer any questions the participant might have, and to show your appreciation for the contribution he or she has made.

Boundaries and Limitations of the CDM

Although the CDM has some major advantages for doing knowledge elicitation, it is not always the best choice. There are situations, task domains, and project constraints where a standard CDM interview is not feasible or is unlikely to yield high quality data. We have encountered two types of conditions that limit the feasibility of the full CDM procedure as a method of CTA.

The first condition that limits our ability to do a full CDM is when there simply are no real experts, or even highly skilled practitioners, to be found. This can happen for a number of reasons. We have encountered domains in which there was only one real practicing "expert." In such cases, getting the person's time is practically impossible. Another possibility is that the job itself may be new, or has undergone a radical transformation in technology and the way work is performed. Practitioners may not have had enough time to build skill in the domain. Or else the nature of the work somehow impedes the development of skilled performance. The domain may be one in which task feedback is difficult to discern, so practitioners are unsure when their actions have actually been successful.

If parts of the task are distributed across time, space, or personnel, then outcomes may become distant from the individual performer. It is very difficult for people to develop significant skill in the absence of clear, specific performance feedback. In a study we conducted several years ago of airport baggage screeners, for example, initial attempts to collect data using CDM were disappointing. Baggage screeners do stop bags that appear suspicious (and we came to understand the basis for their judgment, see Kaempf, Klinger, and Wolf 1994), but they get little if any feedback about how many they miss. And there are many opportunities to miss. On a typical shift in a busy airport, they were screening thousands of bags, often spending less than five seconds per bag. On a more practical level, the job has traditionally had very high turnover rates. As a result, many baggage screeners didn't develop a strong base of experience and robust skills.

A second condition that can limit the usefulness of CDM is one in which participants are unable to generate useful incidents. Combat-like conditions, where people work under severely stressful conditions and handle very high workloads, can create a blur of events that are difficult to recall as discrete cases. For a project on air campaign planning and targeting, we interviewed Air Force personnel who had been deployed to the Persian Gulf during Operation Desert Storm in 1990. Targeting personnel worked long shifts for many weeks prior to and during the conflict. They handled hundreds of targets, working from air campaign plans to acquire needed information for selected targets, briefing pilots, debriefing pilots, and conducting battle damage assessment. They were able to describe many aspects of the targeting task, but they found it extremely difficult to describe an intact case from beginning to end.

In the years since the CDM was first developed we have found that the same principles of incident-based probing can be used flexibly in a wide variety of types of CTA projects. In the sections that follow we describe some applications of CDM interviewing techniques that do not rely on retrospective accounts.

Adaptations of CDM

"Classic" CDM was developed to study decision making in naturalistic settings. The naturalistic decision making (NDM) perspective has widened its field of view, and many NDM researchers, ourselves included, are investigating a range of cognitive phenomena that extend beyond decisions and decision making. (See the discussion of macrocognition in chapter 8.) In parallel, CTA practitioners have expanded, adapted, or altered the CDM technique in a variety of interesting and effective ways. These adaptations retain the emphasis on extracting incident-based data and reliance on people's lived experience as a basis for knowledge elicitation. They greatly expand the utility of the method. Most importantly, they demonstrate the breadth of possibility for application of incident-based knowledge elicitation.

CDM and Here-and-Now Incidents

Observation of experienced people at work is an important activity in CTA. In some disciplines where work is studied, such as the field of cognitive anthropology, observation is the main method. There are a number of ways in which observations and structured interviewing can be combined. In the study of baggage screeners in which CDM was impractical, the researchers developed an approach that relied on observations in airports—standing side by side with screeners and asking them questions about various aspects of what they were looking at and thinking about. Another example is from a project on the mediation of civil (e.g., noncriminal) legal cases (Crandall et al. 1996). The goal of the project was documentation of the cognitive skills involved in dispute resolution and how skilled mediators use their prior cases to help them plan for and carry out a mediation effort. The researchers were fortunate to gain the participation of a leading dispute resolution firm made up of attorneys and former judges. They allowed the researchers to shadow them during actual dispute resolutions.

Mediators typically conduct an initial mediation session with all parties present and then work with the disputing parties in separate rooms. The mediator shuttles back and forth between parties, discussing issues, listening to complaints and grievances, suggesting options, and (ideally) bringing the parties to agreement around a final settlement. Sessions often take several hours, some a full day or longer. In this project, the researchers stayed with the attorneys throughout dispute resolution, moving between rooms with them and eliciting responses to probe questions between the meetings with the individual parties. Sometimes these elicitation opportunities lasted only a few minutes. Sometimes they were as long as twenty minutes. Probes focused on the attorney's view of the mediation at that point in time. At the conclusion of the mediation, the researchers conducted CDM interviews, using the observations and the attorney's responses over the course of the resolution to fill out the attorney's incident account as the basis for additional data collection.

An abbreviated version of the CDM can be used to inform the researcher about a new domain and establish rapport with the participant. The method can also lead to the identification of potential leverage points, a tentative notion of practitioner styles, or tentative ideas about other aspects of cognitive work. The participant is asked to recall a salient recent case and describe his or her goals and activities. A timeline can be constructed, and, using that to anchor the discussion, the participant can be asked about any of a number of things, including information requirements (e.g., What information did you need or use to make this judgment?), mental modeling (e.g., As you went through the process of understanding this situation, how did you understand the problem scenario? Can you draw me a diagram of what it looked like?), and knowledge (e.g., How did this case relate to typical cases you have encountered? How did you use your knowledge of typical patterns?). This abbreviated CDM may be helpful when a full CDM interview is not feasible but the researcher wants to get a feel for some incidents.

CDM and Typical Incidents

Our discussion of the CDM brings with it the notion that a focus on critical decisions is often a good window into cognitive work. But not all CTA methods that are incident-based rely on the study of critical incidents. Some studies cannot rely on the study of critical incidents for the simple reason that not all real-world events comprise critical incidents. Thus, observations or interviews conducted during or immediately after real incidents will not necessarily end up speaking to critical decisions. And in some projects, the notion of "challenging event" doesn't make good sense at all, or it may not work well for the project goals.

Sometimes CTA researchers want to understand how things usually or typically work. In other cases, concerns around memory issues may lead the researcher to go after very recent events, to make sure memory for details is fresh. We have encountered these issues in some of the consumer projects we have conducted. If the product purchase or product use is a frequent event, we may ask for a typical experience of a particular type or we may request the most recent example. Similarly, in a project on physician-patient communication, we asked patients for examples from their most recent doctor visit, rather than a challenging one.

Variations on Use of a Timeline

In some instances, the requirement to elicit a timeline and structure the interview around it simply gets in the way (Militello et al. 2002). Interviewers find it easier and more effective to move directly from the initial incident account to the deepening phase, particularly when the story segments are clear. How do you know when a timeline matters and when it is a frill? The answer is often contained in the domain and the incident itself; a timeline matters when the outcome of the incident depends on time or timing.

An example of a job where time and duration are critical comes from a project on landing signal officers (LSOs), whose job it is to help pilots land planes on the decks of aircraft carriers (Thordsen 1998). Landing an aircraft on a carrier at sea is a difficult, dangerous task. The fact that the landing strip is in motion is only a part of the difficulty. In addition, there is a very small window of opportunity, when the aircraft and the ship are lined up and synchronized and chances for success are optimal. If the window is missed (or doesn't open at all), the pilot must go around again. The LSO has just about forty-five seconds to make the determination to permit landing or to wave a pilot off and require another approach. In order to understand the LSO incidents, the CTA had to yield representations of the task and the LSO's cognitive activities to the second.

In some domains or situations, the outcome may depend not on time but on the particular *sequence* in which events occur relative to other aspects of the situation. In other domains or situations, spatial/geographic elements matter more than time. In a CDM study conducted with Alaskan pilots, researchers used maps to anchor incident segments and decision points rather than an actual timeline (Holbook, personal communication 2002).

Sometimes both sequence and geography matter. This is the case in many military or tactical situations. Linking aspects of a recalled incident to time or distance elements is the best way to ground the story. Another feature to consider is the length and complexity of the incident. Incidents that span several hours or more usually have many decision points, situation shifts, and multiple players. Here, a timeline can be a valuable aid to keeping all the details straight and in proper sequence. What matters is to figure out the structuring mechanism that will best support management of the interview in terms of making sense of the incident, keeping the sequence and details straight, and unpacking the important cognitive elements.

Conducting CDM Over Multiple Sessions

We mentioned earlier that CDM sessions can elicit detailed stories that take time to tell, retell, and deepen. In some domains, and for some types of incidents, fitting the CDM procedures into the standard two or so hours simply does not work. In these cases, it can be productive to divide the CDM into two or more sessions. One approach that works well is to conduct a first session that includes sweeps 1 and 2: incident identification and selection followed by timeline development and verification. The second session includes deepening and "what if" querying. In a three-way split, the first session includes incident identification and selection. During a break, the interviewers transcribe the notes and prepare for the second session. The second session includes recounting the incident and timeline development. Again in the break, interviewers transcribe the notes and prepare for the third session. The third session includes timeline verification and decision-point identification, deepening and "what if" queries. Using this approach, the CDM sessions can be conducted over several days, enabling interviewers to document and absorb details of the complex incident and allowing participants to come to sessions refreshed.

The Knowledge Audit as Incident-Based CTA

The most thoroughly tested and validated adaptation of the CDM concept is the Knowledge Audit method (Hutton and Militello 1996; Hutton, Militello, and Miller 1997; Klein and Militello 2004; Militello and Hutton 1998). The Knowledge Audit examines the nature of the expertise needed to perform work skillfully. It structures an interview around a set of probes covering different aspects of expertise.

The CDM and the Knowledge Audit have sometimes been presented as a contrasting set: complex versus simple, incident-based versus general knowledge, and depth versus breadth of information. Hoffman, Coffey, and Ford (2000) regard the Knowledge Audit as a shortened or truncated CDM. In fact, the two procedures do share points of commonality, but offer distinct views of cognitive phenomena by using different elicitation techniques. The Knowledge Audit poses questions about specific cognitive elements that are characteristic of experts, based on the extensive research literature about expertise (e.g., Chi, Feltovich, and Glaser 1981; Ericsson and Smith 1991; Klein and Hoffman 1993). An example is the item designed to elicit information about perceptual discriminations:

Experts are able to detect cues and see meaningful patterns that less-experienced personnel may miss altogether. Have you had experiences where part of a situation just "popped" out at you, where you noticed things going on that others didn't catch? What is an example?

The Knowledge Audit was developed as a streamlined interview technique, designed for ease of use and accessibility. It is well suited to researchers who are new CTA practitioners. It can also be useful as the very first interview in a project because of the breadth of view it can provide.

The Knowledge Audit covers eight dimensions of expertise:

1. Past and future
2. Big picture
3. Noticing
4. Job smarts
5. Improvising/spotting opportunities
6. Self monitoring
7. Anomalies
8. Equipment difficulties

The purpose of the Knowledge Audit is not to demonstrate the importance of these factors—that is taken as a given. Rather, the purpose is to identify specific skills and perceptible patterns in the context of situations in which they have occurred and the expert's specific strategies for dealing with those situations. The Knowledge Audit is, therefore, useful in the exploration of apprentice-proficient-expert differences. Like the CDM, the Knowledge Audit draws on recall and description of examples. However, it bypasses the CDM requirement to identify and elicit a particular type of critical incident. Instead, the Knowledge Audit provides a structured interview format and set of predefined dimensions for eliciting and collecting examples. Knowledge Audit interviews produce a set of brief stories or minicases that illustrate how expertise plays a role in the particular domain. In a fairly limited time span and with a handful of participants, it is possible to generate a large set of examples organized around a well-defined and systematically applied set of dimensions. Working from the original concept, Knowledge Audits have been developed for use in studying cognitive aspects of team performance (Klein et al. 1999; Militello et al. 1999; Militello et al. 1994), macrocognition (Klein, Ross et al. 2003), and sensemaking (Klein et al. 2002; Klein, Phillips et al. 2003).

Incident-Based CTA with Teams

Many work situations and task functions are carried out by teams.[2] One can gain a very different picture of a work domain by examining the cognitive processes that underlie a team's skilled performance of tasks. Teams process information, make decisions, develop (and lose) situation understanding, detect and solve problems, and make plans (Cooke et al. 2000; Endsley and Jones 2001; Klinger and Thordsen 1998; Salas et al. 1995).

There are a number of team CTA methods currently in use (Klein 1998), and they include incident-based techniques. A version of CDM that has been adapted for use with teams allows the researcher to elicit critical cognitive elements from multiple perspectives on a shared event. A CDM session is conducted with individual team members, who are each asked to describe an incident that is identified for them by the interviewer. The interview proceeds through the four sweeps. The cognitive probes may include questions regarding information sources and targets and aspects of coordination. The outcome is a data set of timelines, cues, goals, expectancies, and information sources, all gathered on the same incident from a variety of perspectives, roles, and functions.

Another adaptation is the team Knowledge Audit. It can be used to elicit aspects of team members' knowledge and skill regarding a specific task or set of tasks, examples, and events in which those skills were required. The team Knowledge Audit probes include: identification of decision makers; mission statement; developing and maintaining the big picture; information management; exposing expertise; team self-monitoring; and adaptability.

Summary

In this chapter, we described one method for using incidents to extract cognitive elements—the Critical Decision Method. We described the procedures for conducting a CDM interview and offered an interviewer's perspective on each of the CDM components. We examined the boundary conditions under which CDM is less likely to be effective, and we described some of the variations and adaptations that have developed to take advantage of the data collection opportunities that real, lived experience offers.

6   CTA Methods and Experiment-Like Tasks

In this chapter we discuss ways to conduct Cognitive Task Analysis (CTA) using experiment-like tasks. Researchers have borrowed or adapted a number of methods from the psychology laboratory to examine cognitive work. One approach to this is to take the practitioner's familiar task, and tinker with it as a scientist might (Hoffman 1987). Therefore, we can sometimes merge the tasks that experimental psychologists use in the academic research laboratory with the familiar tasks in which domain practitioners engage.

- In *constrained processing (CP)* methods, familiar tasks or routines are constrained in some way. The participant may be explicitly instructed to adopt a particular strategy, for example. Conversely, the participants may be confronted with a task that challenges their usual strategy.
- In *limited information (LI)* methods, participants are asked to solve problems given incomplete information.

The challenge for the researcher is to create a task that resembles the familiar task in certain key respects. The task cannot be so different from the familiar task as to make the expert feel uncertain, reticent, or worse, handicapped. If the task deviates too much from the expert's familiar task, it might only inform us about cognition relative to that artificial task—which is itself one of the methodological issues of cognitive psychology (Jenkins 1978; Newell 1973). Compared to what the experts usually do, the ideal CP/LI task needs to yield data that possess ecological relevance, validity, and representativeness. Yet, the task should also involve some sort of control and manipulation (or at least selection) of variables. These aspects of the empirical method are necessary in order for a method to qualify as experimental; or at least be able to disconfirm hypotheses or establish boundary conditions.

Examples

Selective withholding of information can be used to reveal experts' strategies and reasoning sequences in different situations (e.g., Hoffman 1997). Tolcott, Marvin, and
Lehner (1987) had expert Army battlefield intelligence analysts think aloud while reasoning about particular scenarios. On the first presentation of a scenario, the information was limited, but over a series of trials additional information was provided to see if
it would lead to the formation of alternative hypotheses, changes in confidence judgments, etc.

In the "20 Questions" procedure (Grover 1983), the practitioner is provided with little or no information about a particular problem to be solved and must ask the elicitor
for information that is needed to solve the problem. The information that is requested,
along with the order in which it is requested, can provide insights into the participant's problem-solving strategy. One difficulty with this method is that the researcher
needs a very firm understanding of the domain in order to make sense of the expert's
questions and provide meaningful responses on the fly. A way around this is to use two
practitioner-participants, one serving as the participant and the other serving as an
interviewer's assistant. The 20 Questions method has been used successfully in expert
knowledge elicitation (Schweickert et al. 1987; Shadbolt and Burton 1990b).

Many studies in the literatures of expertise studies and applied cognitive psychology and human factors involve combinations of limited information and processing
constraints.

Example 1

This study (from Hoffman 1986, 1987) involved the domain of aerial photo interpretation. In their familiar task, domain experts engage in systematic analysis of aerial photos for evaluation of such things as environmental impact, land use policy, and siting
for large construction projects. The "terrain analysis" task can take days and involves
preparation of a number of map overlay products (soils, underlying rock, vegetation,
drainage, etc.). The process relies on all available information (e.g., topographic maps)
and not just the aerial photos.

In Hoffman's (1987) study, experts were presented with aerial photos but were
allowed only a two-minute inspection period. The experts then had to recall everything they could about the photos and provide their interpretation (e.g., "This region
is an arid climate with shallow soils overlying tilted interbedded sandstone and limestone"). At first, the experts balked at the artificiality, but when encouraged to think
of the task as a game rather than as a test or a challenge to their expertise, they found

the task to be interesting. The task was based on one of the activities involved in the experts' familiar task, but put a severe restriction on both time (constrained processing) and the amount of information available (limited information).

Results from this task revealed the extent to which the experts seem to achieve immediate perceptual understanding of terrain when viewing aerial photos. For instance, after inspecting the tropical imagery for two minutes one expert commented: "If you were to send troops there they would have to be protected from bacterial infections." When asked how he knew that, the expert commented that he could tell from the ponds. The expert could see bacteria in a pond from forty thousand feet? No, but what the expert could see was flat interbedded limestone (in a homogeneous forest, the tree canopy informs about the terrain slope) in a tropical climate. Because the bedrock was flat-lying, the streams led into ponds having no major tributary or distributary for an outlet. Stagnant water in a tropical climate means leguminous water plants, implying that the waterways would be laden with bacteria. This all seems like a long inference chain in retrospect, but in the image inspection task, it seemed more a matter of immediate perception built upon a refined base of perceptual learning.

Example 2

This study (from Hoffman 1998) was intended to reveal the informational cues that expert weather forecasters use in interpreting infrared satellite imagery. Participants were forecasters with the National Weather Service and the UK Meteorology Office. Weather forecasters daily inspect satellite imagery in the form of loops covering a span of time usually on the order of hours. In the task Hoffman devised, a series of five infrared images showing the eastern United States was taken from a loop. The first and second images were separated by a one-hour interval, the second and third by a one-hour interval, the third and fourth by a six-hour interval, and the fourth and fifth by a one-hour interval. The selection of the images was such that the six-hour gap coincided with a merging of two low-pressure centers along a front. Unless the interpreter looked closely at a relatively less-salient weather feature (a small convective cell in the Gulf Coast) the interpreter might misinterpret the weather dynamics and hence become confused about the temporal ordering. In the task, the five images were presented in a random order and the participant's task was to determine the correct ordering. Most of the experts, unlike novices, were able to determine the correct ordering, although experts were not immune to the sequencing misinterpretation.

Following the sequencing task was a surprise recognition task. The participant was presented with a set of images and had to judge whether each image was one of those presented in the ordering task. The recognition set included the originals, plus a few foils. Some foils were images taken from an entirely different weather situation, some were images from the same loop but at different times from those in the original set, and some were visible-light images including ones that were from the same weather event as depicted in the original series. Both novices and experts tended to correctly label the visible-light images as "new," but in different ways. The novices, upon seeing the visible-light images, would focus on literal stimulus features and remark that they looked somehow different from the original set, that the original set (infrared images) seeming pixilated or digitized. Experts, upon seeing a visible-light image, would remark that it was "new" but would add that it showed the same weather as in the acquisition series. "Oh, by the way, this image is of the same weather as the original set."

Both novices and experts found the recognition task to be confusing, but for different reasons. The novices had to base their judgments on their literal understanding of the pictures and would sometimes correctly recognize an image because of literal features (e.g., "I remember seeing a cloud right over Long Island"). What experts remembered was not so much the literal pictures as their understanding of the weather dynamics (fronts, high and low pressure systems, the location of the jet stream, etc.). Hence, they said they recognized some foil stimuli that depicted the same weather dynamics as in the original set. This experiment demonstrated the fact that experts perceive weather dynamics based on complex configurations of cues and do not interpret images based on a literal understanding or set of isolated stimulus features. It revealed the cue configurations suggestive of such things as fronts and the jet stream—all those things that are "invisible" to the novice.

A particular method for studying cognitive work during the conduct of tasks has been widely used in the study of problem solving and the study of expert-novice differences. This is the think-aloud problem solving (TAPS) method.

Think-Aloud Problem Solving

The TAPS task originated in classic research on the psychology of problem solving by Edouard Claparede (1917) and Karl Duncker (1945). In their studies, research participants thought out loud while working puzzles. Task performance could be linked to mental operations, as for example, when a sharp drop in the time it takes the participant to solve a problem follows a trial in which the verbal report suggested that the participant had gained some insight into the problem (see Woodworth 1938). In the recent work on expertise, TAPS has been used widely, and with considerable success, in the study of expert-novice differences. For instance, the classic Chi, Feltovich, and

Glaser (1981) study of physics problems engaged participants (advanced graduate students) in mechanics problems (involving weights, pulleys, etc.) and had them think out loud. Results revealed the breadth and depth of expert knowledge and how they "pre-think" problems by first developing an understanding of problems in terms of the pertinent physical principles. Novices (undergraduates) tended to just dive in and try to determine what equation to solve based on literal problem features. Numerous studies of problem solving using TAPS have successfully charted reasoning sequences (e.g., Bailey and Kay 1987; Claparede 1934; Duncker 1945; Ericsson and Simon 1984; Johnson, Zualkernan, and Garber 1987; Voss, Tyler, and Yengo 1983). Think-aloud problem solving has been used in studies of expertise in such domains as medical diagnostics (e.g., Chi, Glaser, and Farr 1988), computer programming (e.g., Jeffries et al. 1981), and process control (Bainbridge 1979; Umbers and King 1981).

The TAPS Procedure

In their book *Protocol Analysis: Verbal Reports as Data*, Ericsson and Simon (1984) present a detailed discussion of the TAPS method and the associated procedure for analyzing TAPS data. The basics of the method are simple, and can be applied to just about any task. Participants are instructed to speak their thoughts as they work on problems, and do so as if they are "speaking to themselves." They are not to try and plan or to explain what they say. They are not necessarily to talk about their thoughts or feelings (which is introspection) but about the problem (which is task explication) (Ericsson 1996). The instructions include simple examples (e.g., "What is the fourth letter after H?"). In addition to thinking aloud during the task, participants can also be presented probe questions afterward, in a procedure of retrospection.

Questions About TAPS

A question that has been raised about TAPS is whether the task of having to think aloud interferes with reasoning or performance. Based on the literature on skills that have become "automatic," it might be posited that experts will experience problems accessing the tacit knowledge that underlies their performance (cf. Anderson 1982; Fitts and Posner 1967). In fact, introspecting, or attempting to exert explicit conscious control over performance relying on implicit processes (that appear to allow for automatic processing) has been shown to reduce motor performance (Masters 1992). However, Ericsson (1996) argues that experts are likely to be more "in touch" (i.e., have better metacognitive skills such as monitoring) with their performance than novices, inasmuch as they are constantly critiquing their skills so as to improve them and outperform their previous standards or the opposition. Furthermore, thinking aloud about a problem is quite different from thinking about one's thoughts and feelings (Wilson and Schooler 1991).

A second consideration is that in many domains, part of the practitioner's job is to be able to think aloud about the problems he or she is working. Perhaps the clearest example would be the task of the coroner, which includes recording a running mono-log as the autopsy proceeds. Another example is weather forecasting. Forecasters often have to give briefings, explain their forecasts to clients, and explain their forecasts to other forecasters. It is also valuable to look more closely at the cases where performance is supposedly automatized. For example, while world-class gymnasts try to "run on autopilot" during competition, they spend a far greater amount of time in deliberate practice in which they engage in a dialog with their coach during their performance.

A third consideration is whether the TAPS task effectively scaffolds the participant. If you were asked to describe the steps you take in tying your shoelace, and why you take each of the steps, you would certainly have at least some difficulty at it. But that does not mean that the "knowledge" is not there. When we perform a highly practiced task, we may not want to get any better at it, and thus we may not spend time critiquing or deconstructing it in order to do better. Therefore, we can be pretty poor at describing the steps required to do it, and doing TAPS as a simple directive can be disruptive to performance. However, if we're interested in a "speed of shoelace tying" competition, we might begin to deconstruct and critique our method of tying to search for more optimum sequences.

An important aspect of TAPS as a CTA method is the data analysis and representation step.

Protocol Analysis

The data from TAPS consist largely of a record of the participant's verbalizations. The data record is often in the form of an audiotape recording. That has to be transcribed and then coded in some way. The procedure for coding a protocol is referred to as Protocol Analysis (PA). As a general data analysis procedure, PA can be applied to any form of protocol, derived from any of a variety of methods and not just TAPS. For example, a PA might be conducted on a transcript of an audio recording of a structured interview or an expert retrospecting about past experiences. Hoffman (1987) tape-recorded the deliberations of two expert aerial photo interpreters who had encountered a difficult case of radar image interpretation. The case evoked deliberate, pensive prob-lem solving and quite a bit of "detective work." Thus, the transcripts were informative of the experts' refined or specialized reasoning.

In the traditional PA, each and every statement in the protocol is coded according to some sort of a priori scheme that reflects the goal of the research (i.e., the creation of models of reasoning). Hence, the coding categories include expressions of goals, observations, hypotheses, and decisions.

Researchers have devised a number of alternative coding schemes for protocol analysis (see for instance, Cross, Christiaans, and Dorst 1996; Newell 1968; Pressley and Afflerbach 1995; Simon 1979). Without exception, the coding scheme the researcher uses depends on the task domain and the purposes of the analysis. If the study involves the reasoning involved in the control of an industrial process, categories would include, for instance, statements about processes (e.g., catalysis), statements about quantitative relations among process variables, and so on (see Wielinga and Breuker 1985). If the study is about puzzle-solving, categories might include statements about goals, the states of operators, elementary functions, and so on. If the study concerns expert decision making, categories might include noticing informational cues or patterns, hypothesis formation, hypothesis testing, seeking information in service of hypothesis testing, sensemaking, reference to knowledge, procedural rules, inference, and metacognition, and so on.

Statements need not just be categorized with reference to a list of coding categories. The coding scheme might be more complex and might involve subcategories. For instance, statements about operators might be broken down into statements about assigning values to variables, generating values of a variable, testing an equation for variable y versus x, and so on to a fine level of analysis (see Newell 1968).

Also, working backwards from a detailed assignment of each and every statement in a protocol, one can cluster sequences of statements into higher-order functional categories (e.g., a sequence of utterances that all involved a forward search or a means-end analysis, etc.) (see Hayes 1989).

We now present three specific examples of PA coding.

Example 1: Coding for Task Procedures
The first coding scheme we use to illustrate PA analysis shows how PA is shaped by project goals. The abstraction-decomposition analysis scheme evolved out of research on nuclear safety conducted by engineer Jens Rasmussen at the RISØ National Laboratory in Denmark (Rasmussen, Pejtersen, and Schmidt 1990). The researchers developed a scheme that can be used for coding interviews and TAPS sessions. The coding representation shows in a single view both the category of each proposition and how each proposition fits into the participant's strategic reasoning process and goal orientation. This scheme is illustrated in figure 6.1. The rows refer to the levels of abstraction for

Levels of Decomposition → / Levels of Abstraction ↓	Whole System	Subsystem	Component
Goals			
Measures of the goals			
General functions & activities			
Specific functions & activities			
Workspace configuration			

Figure 6.1
The coding scheme of abstraction-decomposition analysis.

Table 6.1
An instantiation of the abstraction-decomposition analysis

	Forecasting Facility	Forecast Duty Officer's Desk	FDO–SAND Workstation
Goals	Understand the weather, carry out standing orders	Forecast the weather	Supports access and analysis of weather data products from various sources
Measures of the goals	Duty section forecast verification statistics	Twenty-four-hour manning by a senior expert	Forecast verification, system downtime, ease of use
General functions & activities	Prepare forecast products and services to clients	Understanding and analysis of weather data	Supports access to satellite images, computer models
Specific functions & activities	Carry out all Standard Operating Procedures	Prepare Terminal Area forecasts, request NEXRAD products	Supports comparison of imagery to lightning network data to locate severe storms
Physical form	The operations floor layout	Workspace layout	CRT location and size, keyboard configuration, desk space

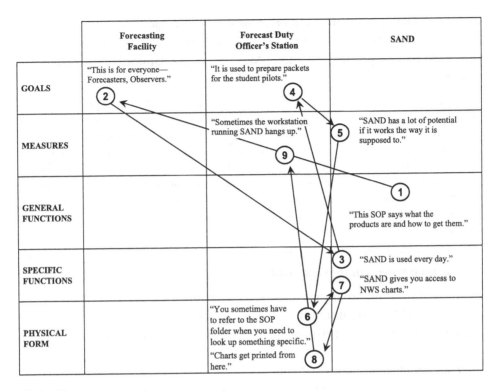

Figure 6.2
Protocol tracing using the abstraction-decomposition analysis.

analyzing the aspect of the work domain that is under investigation. The columns refer to the decomposition of the important functional components.

Table 6.1 shows how this scheme would apply to the domain of weather forecasting (Hoffman, Coffey, and Ford 2000). In particular, it refers to a workstation system for the Forecast Duty Officer called the Satellite, Alphanumerics, NEXRAD, and Difax Display System (SAND).

Once a template for use in a work domain has been diagrammed in such a manner, each of the statements in a specific interview or problem-solving protocol can be assigned to the appropriate cell, resulting in a process tracing that codes the protocol in terms of the work domain. An example appears in figure 6.2. The numbered path depicts the sequence of utterances in an interview in which a forecaster was asked to describe the standard operating procedure involved in the use of the SAND system, supported by probe questions about what makes the system useful and what makes it difficult.

Example 2: Coding of Propositions for a Model of Knowledge

Also from the weather forecasting project, statements from interviews were highlighted that could be incorporated in Concept Maps of domain knowledge (see chapter 4). From the transcript:

...Because usually in summer here we're far enough south that ah...high pressure will dominate, and you'll keep the fronts and that cold polar continental air north of us. Even if the front works its way down, usually by the time it gets here it's a dry front and doesn't have a lot of impact....I also think of ah...I think of tornadoes too. Severe thunderstorms and tornadoes are all part of the summer regime. And again, tornadoes and severe thunderstorms here are not quite as severe as they are inland like we talked about last time, because they need to get far enough in to get the mixing, the shearing and the lift. And that takes a while to develop and unfold. You really don't see that kind of play until this maritime air starts to cross the land-sea interface.

Starting at the beginning of this excerpt, one sees the following propositions about high pressure systems in the Gulf Coast:

(In the Gulf Coast region) high pressure will dominate (in the summer)
(High pressure) keeps fronts north of us (the Gulf Coast region)
(High pressure) keeps cold polar continental air north of us (the Gulf Coast region).

Example 3: Coding for Leverage Points

The weather forecasting project also involved an analysis of standard operating procedures documents. The participant went through each of the Standard Operating Procedures and was probed about each one, as illustrated in table 6.2. The purpose of the coding was to identify leverage points, indicated by the bold text. These are places where a change in the workplace, possibly involving new technology, might improve the work.

The entries in this table are the researcher's notes from the interview, including some paraphrases and synopses of what the participant said, and are not an utterance-for-utterance protocol. This underscores the idea that PA is a general data analysis method even though it is usually associated with the TAPS task.

Coding Verification

In CTA contexts, it is often valuable to have more than one coder conduct the protocol coding task. In some cases it is necessary for demonstrating soundness of the research method and the conclusions drawn from the research. For research in which data from the TAPS task are used to make strong claims about reasoning processes, especially rea-

Table 6.2
Notes from an interview concerning an SOP

Task/Job: Observer updates SIGMETs *

Action sequence: Conducted every hour (valid for two hours)
SIGMET is defined and identified
Change is saved as a JPEG file
Change is sent to the METOC home page and thereby to the Wall of Thunder

What supports the action sequence?	*What is the needed information?*
METOC PC	**How to find the SIGMETS (they come from Kansas**
PowerPoint with an old kludge	**City). Have to plot them—sometimes look up stations**
There is a series of four maps:	**in the station identifier book. Have to know more**
—SE Texas–East Coast	**geography than many observers know.**
—FL, AL, MS	
—VA through GA	
—TX, OK, AR	

What is good or useful about the support and the depiction of needed information?
The SIGMETS themselves are really good—give customers good information at a glance.
The PowerPoint maps are designed for METOC—they have all geographical information and three letter identifiers (e.g., PNS for Pensacola) that are needed. Other forecasting offices have blank maps of the United States.

What about the support or information depiction makes the action sequence difficult?
Too labor-intensive—the whole system is archaic. There is a commercial website that has a Java program with map and red box for SIGMET—you can highlight the box and get text describing the SIGMET. This always seems to be updated.
Limited capability to customize the shapes of the SIGMET areas.
The map cannot move as you are in the act of drawing a SIGMET—you have to change functions and scroll the map with the mouse.
The alphanumerics are hard to see even if you zoom.
The map shown on the CRT is not large enough; details are hard to read.
It is a sectored map—cuts off at Texas.
You sometimes have to hunt for station identifiers—end up searching via "Yahoo." Some stations have several IDs.
Map cannot scroll outside a limited area.
Nothing in the work environment reminds the observer to conduct the task.
NOTE: **The final display of SIGMETs does not support a zoom function. They aren't easy to see on Data Wall.**
The work is often done on the hardcopy map lying on the table—where you can see all the regions and station identifiers in a glance and read all the details. After figuring it out on the hardcopy map the observer inputs it into the computer.

---

*Significant meteorological events

soning models that assert cause-effect relations among mental operations, the assessment of inter-coder reliability of protocol codings is regarded as a critical aspect of the research (see Ericsson and Simon 1993).

In the simplest procedure, two researchers independently code the statements in the protocol and a percentage of agreement is calculated. Researchers typically consider a high rate to be 85 percent or greater agreement among multiple coders (see Hoffman, Crandall, and Shadbolt 1998).

Analysis of multiple codings can show whether the coding scheme is well defined, consistent, and coherent. It is likely that there will be disagreements, even between coders who are practiced and are familiar with the task domain. Disagreements can be useful pointers to ways in which the coding scheme, and the functional categories on which it is based, might be in need of refinement.

In another verification procedure:

1. Two or more researchers, working independently, go over the typed transcript and highlight every statement that can be taken as an instance of one of the coding categories.
2. Each researcher codes each highlighted statement in terms of the coding categories.
3. Each researcher codes the highlighted statements from the *other* researcher's highlightings.
4. Both the highlightings and the codings from the researchers are compared.

A shortcut on this general approach to coding verification is to have multiple coders and a reliability check, and once an agreement rate of 85 percent or more is achieved, all remaining transcripts can be coded by individual researchers.

For some research, the assessment (elaborate or otherwise) of inter-coder reliability may not be necessary. For instance, the identification of leverage points in the analysis of the standard operating procedures in the weather forecasting case study (Hoffman, Coffey, and Ford 2000) did not require an assessment of inter-coder reliability because the leverage points were explicitly elicited from, and were identified *as* leverage points by the domain expert. Furthermore, the coding scheme was simple—a statement either was or was not an expression of a possible leverage point. Likewise, in the protocol analysis of the Concept Mapping interview, the identification of statements that could be used in Concept Maps did not mandate verification that *all possible* statements that could be used in Concept Maps were in fact identified and used. Such an analysis would have actually detracted from the main goal of the analysis, which was to identify propositions that could be used in Concept Maps on particular topics and that were not already in the Concept Maps that had been created.

Effort Considerations

No matter what task is used to generate the protocol, transcription and analysis is very time and labor-intensive (see Burton et al. 1987; Burton et al. 1988; Hoffman 1987). It takes more than an hour for even an experienced transcriber to transcribe each hour of audiotaped TAPS, largely because even the best typist will have to pause and rewind very frequently and cope with many transcription snags (e.g., how do I transcribe hesitations, "um's, "ah's," etc.?). Coding also takes a considerable amount of time, and the validity check (multiple coders, comparison of the codings, resolution of disagreements, etc.) takes even more time.

Both Burton et al. (Burton et al. 1987; Burton et al. 1988) and Hoffman (Hoffman 1987) compared TAPS + PA with other methods of eliciting practitioner knowledge (interviews, sorting tasks, etc.). Both found that TAPS + PA is indeed time-consuming and effortful and yields less information about domain concepts than limited information or constrained processing techniques do. On the other hand, if the analysis focuses just on identifying leverage points or culling information about practitioner reasoning, and not on the coding of each and every statement in the protocol, then PA can be useful in CTA.

In planning to conduct a PA procedure, there are a number of questions the researcher might consider at the outset. These are presented in table 6.3.

Table 6.3
Some questions for consideration when planning a PA procedure

---

What are the purposes?
- If the purpose is to develop reasoning models, then the categorization scheme needs to include such (slippery) categories as "observation," "goal," and "hypothesis."
- If the purpose is to identify leverage points, the categorization scheme can be simple (i.e., highlighting statements that refer to any sort of obstacle to problem solving).

What is the level of analysis?
- Do I cut a coarse grain—"notice," "choose," "act"?
- Or do I cut a fine grain based on a functional analysis of the particular domain? (e.g., "look at data type X," "notice pattern Q," "choose operation Y," "perform procedure Z," and so on).

Do I need to code each and every statement?
- What constitutes a statement? Do I separate them by the hesitations on the recording?
- How do I cope with synonyms, anaphora, and the like?
- Statements are often obviously dependent on previous statements. Context dependence is always a feature of PA because context dependence is always a feature of discourse.

How intensive must the validity check be?
- Indeed, must there be a validity check at all given the purposes of the research?
- Do I need to have independent coders code the protocol and then compare the codings for inter-coder reliability? How many coders—two? three? What rate of agreement is acceptable? How do I cope with the inevitable disagreements?

Variations on the Theme of Experiment-Like Tasks

In deciding whether or not to devise and conduct an experiment-like task, as for any task, the researcher needs to consider methods' strengths and limitations relative to the project goals. Experiment-like tasks may be useful, for example, in probing the specialized sub-domain knowledge or reasoning of experts (Hoffman 1987) or probing specific hypotheses about reasoning or strategies. They may be less useful if it seems impossible to compose a task that possesses ecological representativeness and validity for the participants who are domain practitioners, but also makes absolutely no sense to novices, apprentices, or journeymen.

This particular aspect of CTA methodology reaches into the relations of applied and basic science, the relations of naturalistic cognitive field research to laboratory experimentation (see Hoffman and Deffenbacher 1993). Field study methods such as interviews and observations can elicit rich, high-quantity, broadly based concepts and rich case stories about decision making in a domain and can do so in a relatively brief amount of time. Those methods can help the researcher build models of reasoning and identify leverage points. However, control is usually by selection of variables (participants, cases, etc.).

On the other hand, CP/LI and TAPS tasks seem more like experiments. Both the environment and the stimulus materials can be controlled by manipulation as well as by selection of variables. In many treatments of the scientific method, this is believed to permit disconfirmation of hypotheses about cause-effect relations.

We do not see this as an either-or choice. Cognitive Task Analysis conducted in a field setting can involve control and manipulation of variables, just as laboratory research can involve capturing the "real world." However, the forms of control, selection, and manipulation of variables can be different in a CTA study than in a laboratory setting. Conversely, the ways in which the "real world" is captured in the laboratory result in significant differences from naturalistic studies. Fortunately, we see many ways to use experiment-like manipulations within CTA investigations, as described and illustrated in this chapter. Archived cases for use in an experiment-like task may be easy or tough, routine or unusual. Processing constraints might involve limited time, strategic focus (e.g., "What kinds of mistakes might an apprentice make at this task?"), or interruptions ("What if" probes). Information might be withheld, or it might be provided when requested. It might be bogus. Participants might be deliberately selected for expertise according to a proficiency scale, and so on.

A combination of strengths would be to use what are perceived to be the more "natural" methods (interviews, observations, etc.) to construct a tentative macrocognitive model (of reasoning, cognition, knowledge, etc.), and then to use methods that are believed to involve more control, selection, and manipulation of variables to study the components of that model. For applied researchers, this might mean probing based on the identification of leverage points for software design. For academic researchers, that might involve testing interesting hypotheses experimentally in a laboratory-like setting.

Summary

In this chapter we explored some intersections of laboratory methods and field research methods for conducting CTA. Researchers can tinker with aspects of a familiar task in a variety of ways, thereby eliciting experts' strategies and reasoning, and do so in structured ways that provide empirical leverage. We also discussed think-aloud problem-solving and protocol methods and the use of analytic and representational formats that can reveal important aspects of cognitive process.

7 Analysis and Representation

The time, effort, and resources required to collect Cognitive Task Analysis (CTA) data are a significant investment. Having made that investment, what happens next? How does one go about transforming a collection of notes from interviews and observations into a set of coherent, meaningful findings? For many seasoned CTA practitioners, analysis of CTA data is the best part of the CTA process—exciting, challenging, and enormously fun. It is a process of exploration and discovery, locating what is important in the data set; it is also a process of organizing and structuring those discoveries in order to communicate them well. This chapter provides a look at that process and what is involved in locating important findings within CTA data.

It is not so easy to find guidance on and detailed description of data handling techniques or on the analysis and representation of CTA data. The knowledge elicitation phase of CTA has received considerably more attention in the literature on CTA methods. To some extent, the emphasis on data collection makes good sense. Data that have been gathered haphazardly, with little thought or planning, are unlikely to produce interesting and important insights or support meaningful, useful applications.

But the flip side of the coin is equally true: a CTA project that has had thoughtful preparation, access to great subject-matter experts (SMEs), and produced rich, varied data may fall short of its goals if, in the analysis phase, the project team fails to take full advantage of the data.

One reason this can happen is that many CTA practitioners have had little training in qualitative research methods. In the fields of psychology and human factors at least, students are typically taught to conduct the analysis phase of research as a series of pre-set plays, using statistical analysis packages. Computerized analysis tools have enormously reduced the time and complexity of using quantitative approaches to explore large data sets.

However, many CTA methods generate data that do not fit easily into standard statistical approaches. A highly structured analysis process that produces quantified data can be attractive, particularly for people trained to consider traditional statistical methods as the only legitimate path to locating what is important in a data set.

However, sometimes the whole really is greater than the sum of its parts. Quantification typically means stripping a body of data of its contextual links and decomposing it in order to assign numerical values. In the process, the research team may sacrifice meaning, particularly larger meanings about complex cognitive processes, macrocognition, and cognition in context (see chapter 8). Quantitative approaches offer important tools, but so do qualitative methods.

The approach to analysis of CTA data that we have found most worthwhile falls in the middle of the analytic spectrum, drawing on both quantitative and qualitative analytic techniques. For example, we might rely on having multiple coders make judgments of weather forecasters' reasoning styles and also conduct statistical analyses of the "hit rate" of the forecasters' weather predictions. The goal in analyzing any CTA data set should be to make full use of the richness of the data. Because CTA projects often have an exploratory component, it is important to be open to possibilities or opportunities to analyze the data in ways that allow new insights and relationships to emerge. This typically involves *qualitative analytic techniques*.

For many people, that means developing knowledge and skill around qualitative methods to balance and broaden the training and education in quantitative analysis they already have. Qualitative methodology is not a standard part of the psychology curriculum in most universities, nor has it found a place in human factors or cognitive science programs, at least in the United States. Those coming to CTA practice from any of these fields are likely to have minimal preparation for analyzing qualitative data at best. That's not to say that the information isn't out there. There are strong qualitative research traditions in the fields of sociology, anthropology, and education. Good textbooks and resource materials (e.g., Creswell 2003; Miles and Huberman 1994) are readily available, and there is active debate about the relative strengths and drawbacks of qualitative and quantitative approaches (see Silverman 2001 for an excellent discussion of this topic).

CTA practitioners can find a great deal of useful information and guidance in these sources. For example, there are excellent descriptions of specific data structuring techniques (e.g., Miles and Huberman 1994), approaches to thematic analysis (Glaser and Strauss 1967; Strauss and Corbin 1997), and useful discussions of approaches to evaluating reliability and validity (Creswell 2003; Hammersley 1992). Nonetheless, the issues they explore and the examples they offer are typically about topics that do not have a cognitive focus, such as analysis of social process or attitudes surrounding terminal illness. The CTA practitioner is faced with the task of selecting what seems most appropriate, figuring out how to adapt it to the cognitive arena, and merging it with techniques and approaches derived from quantitative methodologies.

Familiarity with a range of qualitative analysis methods is obviously important. At least as important is the researcher's stance—that is, how the person approaches the task. The analysis of qualitative CTA data requires the analyst to engage in a discovery process, to be thoughtful, curious, and open-minded in pursuing the data's full meaning. That curiosity and open-mindedness plays out in questions such as, "What is surprising here? What did I expect to find that isn't evident in this data? What are the data telling me that I don't already know? How does this data help me to think differently about issue X?" These are powerful questions that will point to aspects of the data that are not so obvious, that go below surface features. Let's suppose that the data contain responses to the question, "How many times in a shift do you do task A?" One approach to analyzing the data might be to tabulate all the answers. The information provided by doing so may be important, but it is also obvious information. In contrast, an analysis that poses the question, "What makes task A occur more often on some shifts, or for some operators, than for others?" digs below the surface to create a new level of understanding and insight. An active, discovery-oriented approach to analysis will allow the analyst to peel back the surface layer of the data and explore what's beneath.

The Analysis Process

There are a number of ways of carrying out an analysis. Primary among them are what the project is about and what the project team must create, represent, document, and/ or deliver at project end. An analysis conducted in support of storyboards and interface concepts is likely to have some different components and representations than one devoted to developing training scenarios. Nevertheless, across a variety of types of products and outcomes there are certain steps, or phases, that most successful analyses pass through.

A description of those phases and the relevant tasks contained in each provides a roadmap for analysis of CTA data, and this is what we describe in this chapter. The process we describe is primarily qualitative, but not entirely so. The issue is not about choosing qualitative versus quantitative methods, but how to gain the best and fullest understanding of the data that the project has produced.

The following sections provide a description of the analysis process as it commonly occurs in our own CTA projects. We continue our case study approach, examining the processes and procedures that have been most useful and effective in our work and offering them as an initial road map. Clearly, the field would benefit from more consistent and detailed explanations of how data evolves to findings and application recommendations. The sections that follow are a start in that direction.

Phases of Analysis

Analysis can be parsed a number of different ways. We have found it useful to approach analysis as a process involving the following phases:

- Preparation
- Data structuring
- Discovering meaning
- Representing findings

   Although we will discuss them as discrete phases, in practice they flow one into the other and interweave (see figure 7.1).

   A particular portion of the data or phase of analysis may suggest questions and issues that lead back to a prior step and a return to that activity. Similarly, although we have characterized knowledge representation as a discrete step in the overall CTA process, the reality is that representations are created throughout data preparation and data handling. Table 7.1 presents an overview of the analysis stages and the goals, tasks, and potential procedures involved in each.

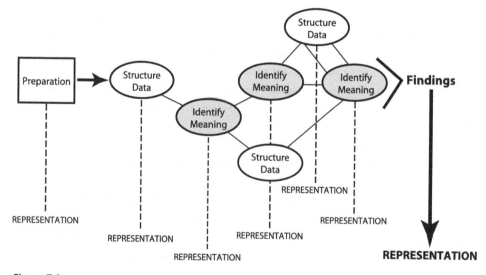

Figure 7.1
The analysis and representation process.

Table 7.1
Processes and procedures for analyzing CTA data

Phase	Goal	Tasks	Procedures
Preparation	Move from informal, intuitive to a structured process of inquiry	Prepare interview and observation data, prepare analysis team	Data records completed and reviewed Project issues and questions reviewed Plan first data sweeps
Structure data	Decompose data into discrete elements; examine pieces and parts; check reliability	Data immersion: identify elements and segments, pull the puzzle apart, examine pieces and parts	Lists Sorts Intervals Coding Cataloguing Frequency counts Descriptive statistics
Discover meaning	Identify central questions, issues, emergent threads of meaning	Structure, integrate, contrast, and compare	Cue sets Categories Patterns and themes Gaps and discrepancies Rankings Ratings Group contrasts Statistical tests
Identify/represent key findings	Make meaning visible; bring into view the story contained in the data	Communicate, display, illustrate, and represent	Stories and incident accounts Charts, graphs, and tables Storyboards Time and event lines Concept maps

Preparing for Analysis

There are a number of activities to accomplish in preparation for moving into analysis. Completing these tasks ahead of time will allow the analysis to proceed more efficiently and effectively.

Preparing the Data

The overall goal here is to evaluate the completeness and accuracy of the full data set, and to plug any holes that are found. For example, have data text files been created from notes or audiotapes for all interviews? Similarly, have records of observational data been created, if observations were part of the data collection? Part of this "final data check" is to ensure that all files have a consistent format so that the project team doesn't spend time puzzling about what it is looking at or where particular information can be found in the data record. Each data record should be clearly labeled with complete identifying information (e.g., interview site, participant name, interviewers, and date of data collection). Any artifacts obtained during data collection—photographs of the work site, sketches and drawings, maps, screen dumps of interfaces, and technology descriptions—should be gathered and labeled. If you have chosen to have transcripts prepared, those should be obtained and reviewed for accuracy (transcribers can have difficulty with acronyms and the vernacular of a domain; we often have to edit transcripts before they are ready to use). Any video or audiotapes should be catalogued and reviewed, if necessary.

Quality-Control Checks

A critical step in preparing interview data for analysis is checking each data record for accuracy. Optimally, this is done at the time the data record is created, soon after the interview has occurred. Because we work in interview teams, we have two data collectors and two sets of notes for each interview session.[1] Our own practice has been to have the interviewer who has served as "Second" during that interview write the notes up in the form of a text file (see chapter 5). This makes sense, because the Second is likely to have taken more extensive notes than the Lead interviewer (who is busy leading the interview). In any case, one of the interviewers agrees to take on the task of creating the text data record for that interview. When that person is finished, the text file is handed off to the interview partner, who reviews the text file against his or her own notes from the interview, filling in gaps, correcting minor errors, and checking and verifying the content of the data record. Occasionally, the two interviewers will find a substantive difference in their notes and in their recollection of the interview. To resolve the discrepancy, they may review the audiotape; if they don't reach resolution that way, they may contact the interviewee by phone or email and solicit additional information. Some CTA practitioners take an additional quality control step and ask the participant to review the interview record. However, this can add considerable time to the data preparation step. At the end of this process, the interview team has created a data record for that interview that is checked and reviewed against an additional set of notes, and other sources if necessary, to verify its accuracy and completeness.

Preparing the Analysis Team

The informal conversations, notes made in airports at the end of an interview trip, and other such thoughts and ideas need to be gathered up. This is the time to review documentation from early in the project: framing questions, goal descriptions and notes from kickoff meetings, the kernel sentence developed to describe the project (see chapter 3)—all should be recovered and reviewed by the project leader and analysis team.

The key question at this point is, "What questions will we ask of the data, and in what way?" The data analysis team needs to settle on an approach, and what to focus on, as they begin the analysis process. Will everyone read all the data records, write down any questions and thoughts he or she has, and reconvene for a first "rough cut" assessment? Will the team divide up assignments, pursue different sets of questions or issues, and report back to the larger group? Will the team review a subset of the data in depth as a way of deciding how to approach the larger data set? There are myriad possibilities. The point is for the team to arrive at an explicit process and set of questions with which to systematically approach the data. The analysis process moves from an informal, intuitive sense of the data to a structured and systematic analysis process.

Documenting the Analysis Process

Be prepared to document the analysis process, decisions, and interim products as the analysis proceeds. It can be a challenge to keep track of the process as it unfolds. Rather like hikers marking the trail to keep track of the route they used, it is important to be able to describe later on what you did and how you did it. Clients often want to know the basis for a general finding, or how applications, concepts, and recommendations are grounded in the raw data. An audit trail that links raw data to eventual outcomes is a critical source of credibility.

With the data and the analysis team ready to go, the next phase of analysis is one of diving into the details of the data and examining its separate elements.

Structuring the Data

Settling into the analysis, the project team members may wonder how to get their arms around the data set. People new to performing CTA may stall out at this point. They sometimes feel overwhelmed at the sheer amount of data and unsure about imposing a structure on it. With experience, people come to relish this part of analysis. It is a bit like going on a trip to a place you've never been before: you don't know exactly how things will unfold, but clearly it will be an adventure! Part of the adventure is learning and figuring things out along the way.

Data structuring is about pulling the data apart and decomposing it in order to organize it in different ways and begin to understand the patterns within it. Later on in the analysis process, there will be time and opportunity to look for themes and to see how relationships generalize across the data set. Getting to that step means first identifying data elements—the pieces and parts—of the data set.

Some people think that data analysis is like solving a puzzle. One way to begin to solve the puzzle is to identify details and discrete elements and to sort them into classes and categories. With a complicated jigsaw puzzle, for example, you might spread all the pieces out, make sure they are all right side up, and then you might begin to sort them by color, shape, or size. Some consistencies and simple relationships may begin to come into view. In order for the task of solving the puzzle to be anything other than random, you must look at individual elements, sort and organize them, and take the time to see what you have to work with.

In the data-structuring phase of analysis, people typically work with individual data records and individual interviews. As a way to home in on the cognitive elements within the data, it can be helpful to begin by reading through the data set, or at least a portion of it, and asking:

• Where is this person's attention? What are they paying attention to, and what are they ignoring?
• What senses are they using? What are they looking at, listening for, touching, smelling?
• What are they thinking about? What are they wondering about, what are they worried about, what are they certain about?
• What information are they seeking, and from what sources?

Reading through the data record with these questions in mind helps us bring the cognition and the context into view and begin thinking about the data in terms of the project's particular questions and issues. It provides an initial orientation to the data set and prepares the analysis team for the data structuring tasks ahead.

Data structuring requires systematically working through the data set and noting any content that pertains to the identified issues and topics. There are many ways to do this: by classifying or cataloguing specific content, making lists, sorting data elements into categories, identifying and marking off critical intervals within each incident, or counting the instances of occurrence of various factors. You could do all of these. Here is an example of an incident account from a project conducted with NICU nurses (Crandall and Getchell-Reiter 1993; also see chapter 3). Reading through it, think about what might be important to "pull out" of the incident account as part of an initial data-structuring activity.

Example 7.1
A Baby in Crisis

I was working with a 30-weeker who was probably about two weeks old at the time. She wasn't my primary, but I had worked with her enough to know when I came on shift one day that something was wrong. All of the data we collected—X-rays, labs—

indicated the baby was okay. But it was my gut feeling that this baby was septic. She just didn't look good. Her color was off from normal, and she was a little floppy. She had sick eyes; they were wimpy looking. But there was nothing clinically that you could point your finger to. Her CBC was normal, she had negative cultures, and she was tolerating feeds. The doctors made rounds on the baby early in the morning and had made routine orders. I chased them down after rounds and talked to them, "I don't have anything really to go on, but I think this baby is getting sick." They told me to keep a close eye on her.

She looked worse as the day went on, but there was nothing obvious that someone who wasn't familiar with her would notice. There was just a subtle gradual deterioration. She had a couple of bradys[2] at 2:00, which was unusual for her. But she was still tolerating her feedings, her abdominal girth was normal, and her vital signs were fine. She really bottomed out that evening, had multiple episodes of bradys and apnea. Within six hours she was on a ventilator. It was really kind of upsetting because I knew it was coming. Maybe if we had intervened sooner she might not have gotten as sick as she did. The lab work for sepsis came back positive, and with antibiotics she came back around. But this was a real setback for her.

In an initial pass through the data, some of the categories that stood out as important to note in this incident account are listed in table 7.2.

Some of the cognitive elements that occur in this incident account include:

- Expectancies
- Decision making
- Problem detection
- Perceptual cues
- Spotting anomalies

In reading through the incident accounts in a data set, the next account may be very different in terms of the categories that "pop out" and the cognitive elements the incident contains, or there may be appreciable overlap. At this point the task is not to try to make things fit, but to allow the data to inform and to explore what it has to say. At the completion of an initial pass through the data, the analysis team is likely to have noticed some things they hadn't thought of before and have come up with several new questions as well. That is one reason that multiple data passes are important. Each pass through the data offers a chance to gain new insights, to evaluate and prioritize new questions in view of the overall goals of the project, and to examine them systematically in subsequent passes through the data.[3]

Table 7.2

A baby in crisis—incident data

Type of Information	Incident Account Content
Patient information	"30-weeker" (moderately premature) "2 weeks old at the time"
Prior knowledge of patient	"Not my primary" "Worked with her enough to know..." "Someone not familiar with her would [not] notice"
Intuition	"Something was wrong" "My gut feeling was..." "I think this baby is getting sick" "...I knew it was coming"
Action	"Chased the doctors down after rounds and talked to them"
Emotional impact	"It was really kind of upsetting"
Cues → sepsis	Color was off, floppy (muscle tone), sick eyes, bradys, apnea
Cues → not sepsis	Normal CBC, negative cultures, tolerating feeds, abdominal girth OK, vital signs OK
Time element	"Subtle gradual deterioration" "Maybe if we had intervened sooner..."

Example 7.1 is from a Critical Decision Method (CDM) study, so that the basic unit of analysis is each incident account. However, the process we are describing holds regardless of the particular knowledge elicitation method used. For example, the analysis team might examine concept maps for similarities in types of "node-link-node" relationships or look for branches that suggest common elements. In the Knowledge Audit, the specific probes provide an initial category structure that can be useful for data structuring activities. If the data set contains observational data along with interviews, those records can be examined in the same way we have described for exploring incident-based data.

At this point in the process, it may be useful to work with hard copies of the data, marking and itemizing the content in different ways. There is no way to do this task incorrectly. It welcomes divergent thinking. Explore different ways to segment, sort, categorize, and code the data. The main consideration is good time management, for this is one of the points where it's very easy to become captured by the details of the data and to lose track of the overall goals and schedule.

The goal of data structuring is to examine the data as a collection of discrete elements and to get a sense of where there are useful and interesting connections. In the next phase of analysis, the task is to examine the various linkages, to order and organize them into larger sets and categories, and to locate the threads of meaning that exist in the data set.

Discovering Meaning

There is a point of transition that occurs in the analysis process, as the focus moves from examining individual data records to more general characterization of the data set as a whole. The central question is: What is the story contained in this data?

The central task in this phase of analysis is to locate the significant findings and insights contained in the data. This happens by systematically examining whether concepts and relationships noticed in individual interviews and subsets of the data are consistently present across the larger data set. Relevant activities might include:

- Integrating and synthesizing data elements by organizing lists, categories, and data codes into more general units, for example, tables of difficult decisions and the cues and strategies consistently associated with them.
- Describing regularities in the data by identifying patterns, themes, and cue sets, for example, developing inventories of critical cues.
- Identifying what is missing; locating interesting gaps and discrepancies by contrasting and comparing various segments or subsets of the data, for example, finding that critical information is consistently shared in one work team but appears to be altogether absent in another.
- Examining group similarities and differences, for example, contrasting experts and novices, different marketing segments, or work settings.
- Performing statistical analyses to empirically examine clusters, group differences, associations, and other types of relationships within the data.[4]

The task of discovering meaning will certainly involve finding a "second level" of data structure, one that allows identification of more general findings and interpretations that cut across individual data records. All these activities will involve creating representations[5] of the data. As you begin to work at a more general level, representations become increasingly significant as an analytic tool. They are important not only for depicting relationships and insights, but also for revealing them and for helping the analysis team locate the key relationships contained in the data.

Table 7.3 provides an example of a second-level analysis of data obtained from the project with NICU nurses we described in chapter 3. The goal of the project was to explore nurses' intuition, to see what lay behind skilled nurses' intuitive "click" or gut feeling about their patients. Data collection had yielded a number of incident reports

Table 7.3

Critical cue inventory: Sepsis in premature infants

Cue	Descriptors
Color changes	Early in onset: pale, washed–out skin tones, most noticeable in extremities; underlying gray tinge to the skin. As infection grows: gray skin tone becomes marked. Some nurses describe the color as green-gray, others as yellow-gray. The "gray" descriptor is always present.
Respiration and heart rate	Episodes of apnea (As) and/or bradycardia (Bs) become more frequent; an accelerating pattern of As and Bs is a significant marker.
Lethargy	Infant is less alert, sleepy, listless. Muscle tone is limp, floppy, flaccid.
Unresponsiveness	Decreased reactivity to stimulation.
Feeding abnormalities	Abdominal distention, increased residuals.

of a particular type of event: the early signs of sepsis, or systemic infection, in premature babies.

The incident account in example 7.1 is one of those. We had not sought incident accounts specifically about sepsis, and it was not our intent at the beginning of the project to make sepsis the focal point of the analysis. But that is precisely where the data led. For example, in every interview that involved the possibility of sepsis, these highly skilled nurses had mentioned changes in an infant's coloring as an indicator that something might be wrong. They described their growing concern as a baby's skin tone changed from pink to pale to dingy gray. The reports were strikingly similar across incident accounts. The data contained consistent sets of cues and patterns for other indicators as well. By linking the specific descriptors to cue categories, a compelling picture emerged from the data. That picture allowed us to understand what the nurses were seeing and responding to and how they knew that a problem was developing: increasing paleness, heartbeat that is losing its rhythm and is beginning to miss, breathing problems, loss of alertness, limp muscle tone, listlessness, and food that isn't digesting. Any one of the indicators would not be a cause for alarm. But taken together they signaled that a baby's entire system was in the process of shutting down.

In this phase of analysis, the goal is to identify central questions, issues, and themes contained in the data and to follow threads of meaning as they emerge. It is useful during this phase of analysis to make periodic checks that the team is staying on track and that the types of analyses and representations being generated are appropriately focused. The expansive, divergent exploration of the CTA data we described earlier gives way to a more convergent, purposeful examination that is increasingly linked to project goals and eventual application of the findings.

In most projects we find it useful and interesting to explore a range of types of data representational formats, including many that rely on color, form, and graphical techniques. In the section that follows, we will explore the data representation phase of CTA.

Representing and Communicating Findings

The analysis of CTA results culminates in the task of representing and communicating the significant findings and central meanings identified so far. Of course, by this stage the project team has a variety of types of data representations (and may in fact be awash in them) that have been created over the course of the analysis. Even the text files created from interview notes are a form of representation (of the interview itself, and of handwritten notes taken during the interview). These early representations tend to be "data driven" in the sense that they are based on individual data records; they are intended as internal communication tools for use by the analysis team.

As the analysis moves into later phases, representations take on a distinctly different look and feel. They are not as tightly tied to specific data records. As the team begins to integrate and synthesize the data, representations become more meaning-driven; they are intended to represent findings of the project and communicate your findings to a variety of people and groups external to the analysis team: users, clients, other researchers. The goal is to make meaning visible, to bring into view the story contained in the data.

It's easy to find examples of data representations. In fact, they are so commonplace that it's possible to forget they are part and parcel of the CTA process. This disconnect happens partially because they are typically identified by their specific content and/or use (e.g., concept maps, storyboards, training scenarios, charts, figures, graphs). And although examples abound, the processes used to develop them are rarely described. Nor is there much guidance available about what to use when, or how to best communicate a particular type of data.[6]

A review of representations from our project archives suggest the taxonomy of representational formats presented in table 7.4. Descriptions of each of the categories, along with examples, appear in the sections that follow.

Narrative Formats

Narrative formats capture the richness of detail in an incident account. Incident accounts can be used to highlight cognitive aspects of performance and to reveal the

Table 7.4

Types of representations

Type	Use
Narrative formats	Depict specific incidents; highlight cognitive aspects while retaining chronology and context
Chronologies	Represent temporal events and sequences of cognitive processes
Data organizers	Compare data elements across specific categories
Process diagrams (cognitive)	Illustrate cognitive elements and their flow over time, type of event, or task
Process diagrams (task/action)	Illustrate work process, task, or set of actions along with cognitive performance elements
Concept maps	Provide graphical representations of knowledge structure within a specific task or domain

"story behind the story." Because they retain chronology and the context, they convey a person's lived experience and how a particular event appeared from that person's perspective. Figure 7.2 is an example of narrative representation, using the incident accounts from the NICU nursing study described earlier (this particular incident was not about sepsis, but about working with infants born to drug-addicted mothers). The left column of figure 7.2 contains a portion of the incident account as it was initially told; the right column shows information that was elicited later in the interview in response to probes designed to reveal nursing expertise. The side-by-side presentation is useful for several reasons. It highlights specific elements of expertise, which may be particularly useful in training or knowledge sharing applications; it also reveals what is beneath the surface of the initial account. When nurses (or other experts) see this representation, it helps them realize how much they may not be saying, and why skilled knowledge elicitation may be necessary. It illustrates the point that skilled personnel may have a wealth of information that does not make its way into standard stories.

Chronologies

Chronologies depict sequences of events. They provide a means for representing how a context changes and how time impacts cognitive aspects of performance. They can also be useful for displaying multiple perspectives on an event, such as the one depicted in figure 7.3.

The event depicted in figure 7.3 is based on observations and interviews conducted during an emergency response drill at a nuclear power facility (Klinger and Klein

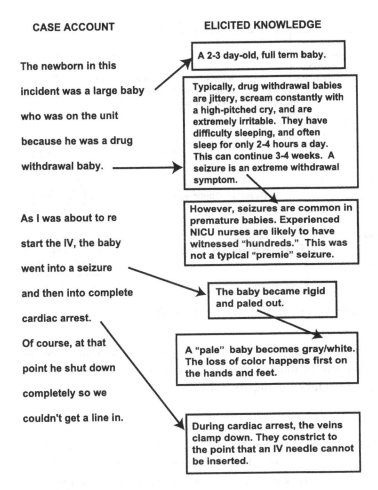

CASE ACCOUNT   ELICITED KNOWLEDGE

The newborn in this

incident was a large baby

who was on the unit

because he was a drug

withdrawal baby.

A 2-3 day-old, full term baby.

Typically, drug withdrawal babies are jittery, scream constantly with a high-pitched cry, and are extremely irritable. They have difficulty sleeping, and often sleep for only 2-4 hours a day. This can continue 3-4 weeks. A seizure is an extreme withdrawal symptom.

As I was about to re

start the IV, the baby

went into a seizure

and then into complete

cardiac arrest.

Of course, at that

point he shut down

completely so we

couldn't get a line in.

However, seizures are common in premature babies. Experienced NICU nurses are likely to have witnessed "hundreds." This was not a typical "premie" seizure.

The baby became rigid and paled out.

A "pale" baby becomes gray/white. The loss of color happens first on the hands and feet.

During cardiac arrest, the veins clamp down. They constrict to the point that an IV needle cannot be inserted.

Figure 7.2
Example of narrative representation format.

1999). As part of the exercise, the emergency response team was told that a fire had occurred in a part of the plant. The emergency director asked a team of engineers to determine the potential impact of the fire and what to do about it. They spent over an hour developing contingency plans. Meanwhile, the emergency director learned that the fire had been extinguished and went on to other tasks. The representation in figure 7.3 shows the event as it developed from the dual perspectives of the emergency director and the engineering crew. It illustrates communication and coordination difficulties that occurred throughout the exercise and that are commonplace in many

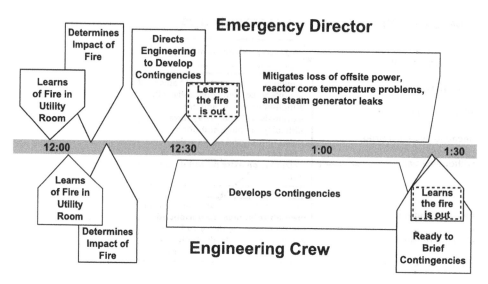

Figure 7.3
Example of a chronology representation of an emergency response drill.

emergency response organizations. Chronologies and timelines are powerful representational tools because they provide multiple views and convey the complexities of incidents and how the element of time can impact events.

Data Organizers

Data organizers are representation formats that provide comparison of data elements across categories. They can be enormously helpful as tools for synthesizing and integrating data—for bringing a variety of data drawn from a single data record into a single representation, or for combining data from multiple data records into a common format. The critical cue inventory shown in table 7.3 is an example of a data organizer. Table 7.5 is an example of a different type of data organizer, a decision requirements table (DRT).

The DRT in table 7.5 is from a project on civil law mediators (Crandall et al. 1996). It presents the typical phases of mediation along with the difficult decisions, strategies, and novice errors that our analysis of the CTA data identified.

The DRT is a representational format that consolidates data from multiple sources into a description of the key decisions and how they are made. The DRT organizes cues and information as well as strategies and practices that surround performance of a given task, documents specific difficulties or challenges associated with its perfor-

Table 7.5
Sample decision requirements table: civil law mediators

Mediation Phase	Decision Challenge	Cue/ Information	Strategy or Practice	Novice Traps
Joint session	Move parties away from adversarial stance; present goal of reaching a settlement both sides can live with.	Body language; tension level.	Tell disputants the goal is not to decide right and wrong; ask them to imagine themselves one month from today: are they comfortable with their conduct?	Fail to inform disputants about mediation process; ignore tension.
Joint session	Elicit personal and relationship issues.	Rely on clients; elicit their story.	Identify key information; do not try to iron out details.	Winner/loser focus; who is more right?
Caucus	Present realistic view of the case.	Client has tunnel vision; no sense of the other side of the case.	Give the hard news about their side of the case; expose client to what can go wrong in court as well as what can go right.	Avoid distress the hard news may cause.
Settlement	Move disputants towards discussion of the dollars.	Ask: "How flexible are you? What standard are you using to judge the right amount?"	Make a number of passes at defining the acceptable range; figure out how the client will respond to a given number, and when sense of that number is clear, cut to the chase.	Suggest dollar amounts without a sense of where the client's range is; too wide a discrepancy can cause disputants to abandon attempts at settlement.

mance, and identifies potential pitfalls and errors. It is a particularly useful tool because it lends itself to organizing information at a variety of levels of synthesis and integration. Decision requirements tables can be used to examine data from individual data records and to combine data across individual data records to reflect general findings.

Data organizers reveal overall patterns, along with the specific indicators and trends that make up a pattern. Their flexible formats can be used to convey different aspects of the data and to indicate how key elements of the context are linked to cognitive performance elements.

Process Diagrams

Process diagrams can be used to represent cognitive performance elements and associated tasks, events, contexts, and action sequences. Process diagrams are useful for depicting cognition in action. Figure 7.4 is a representation of the assessment and problem-solving skills of personnel who are responsible for certifying the safety of commercial aviation (Hutton et al. 2003).

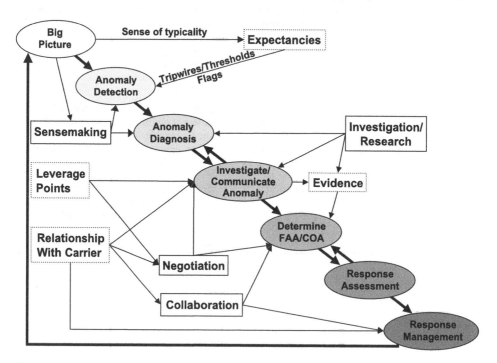

Figure 7.4
Certification management: aviation management oversight.

In this particular representation, the major tasks involved in safety oversight are depicted in the cascade of key functions. These are functions widely recognized in aviation safety as part of certification management and the safety oversight process. A variety of additional cognitive processes, management issues, and other factors were identified in the CTA and are depicted here as well. These are critical cognitive tasks that are central to the oversight of air carrier safety. Process representations can provide insight into the multiple layers of cognitive activity that are involved in a task that on its surface may appear well structured, steplike, and sequential.

Concept Maps

Concept maps are graphical depictions of the knowledge structure in a specific task or work domain. Chapter 4 presents an in-depth discussion of concept maps, along with a number of examples of applications of concept mapping.

Although all of these formats can serve as the basis for formal representations to communicate with external audiences and users, they are also useful for less formal, data-driven representations at earlier stages of analysis. Project findings are seldom captured by a single representation or type of format. Clearly, the representational formats described here capture different elements of cognition and context. It may require several types of representations developed on the same data set to convey key findings and significant insights adequately.

Practical Issues: Factors That Impact Quality of the Data Analysis

The efficiency and effectiveness of data analysis can vary considerably. Some projects bloom with unexpected insights. Occasionally, a promising project yields less than expected. There are aspects of the analysis process that can make a significant difference in the quality of findings and subsequent application. These factors are leverage points that can make the difference between skimming the surface of the CTA data and taking full advantage of its richness and complexity.

Multiple Passes

One factor that can impact the quality of data analysis is having the time and resources (and willingness) to make multiple passes through the data. A data pass is an examination of the full data set (for example, all interview records) with a specific set of questions in mind. It is part of the systematic aspect of handling the data, to examine the entire data record with a particular focus in mind.

It is simply not possible to mine CTA data well in one data pass. How many times you work through the data set in this systematic fashion will depend on the time and resources you have and perhaps the size of the data set. Whatever those limits are, they should allow for a minimum of two to three data passes. What can you gain from multiple passes, other than increasing familiarity with the data? You can examine the data from different perspectives, segment issues, and project questions. You can allow each pass through the data to inform the next, so that insights and questions that the data reveal can be tagged for subsequent, systematic attention.

The systematic treatment of a set of questions is part of the analysis process. But it's worth noting that the analysis process actually begins with the very first interview. Our data collection typically happens out in the field, and we may spend several days conducting interviews. We work in interview teams, and as we head to the airport and wait for our flights home we invariably start processing the interviews—talking over what we heard, what we saw, what we expected to get that we didn't, what the biggest surprises were. This is initial analysis, and the insights gained on the heels of a set of interviews can be important and informative later on. As results emerge, the CTA method is adapted and fine-tuned. Taking the time to jot down a few notes and capture these first thoughts and impressions can pay off later on, when you are ready to dive into the data in earnest.[7]

Staying on Track

A second factor that can affect the quality of the data analysis is keeping an eye on the project goals and the client's needs. This may seem so obvious it is hardly worth writing about. Yet becoming sidetracked in the data is one of the most common problems that project teams encounter. Data analysis is an activity that takes place down in the weeds. In order to do justice to the data, the weeds are where you need to be, for at least some period of the analysis. Once down there, it is very easy to lose track of the big picture. It is also easy to become convinced that our view of the data is conclusive, meaningful, and true. The very act of analyzing the data can get us off track and down a side path. Given these distractions, it is important to find ways to regain focus and periodically check our thinking as we move through the analysis process. Some suggestions:

Work in a Team   Expose ideas about the data to people who also know the data well. We may be very certain about a pattern in the data, but our analysis partner may say, "We heard that from the SMEs in the first set of interviews, from Facility A. I've been working with the interviews from Facility B, and it's not coming through in that part of the data nearly as clearly."

Periodically Review Written Goals for the Project   The kernel statement the project team has created about the project and the work carried out around framing questions (chapter 3) will be of service here. It can be helpful to be reminded periodically of goals. An example is a project conducted for the Air Force Weather Service. It was a large project, with several large and interconnected questions about forecasting expertise, technology, training, and personnel selection. We had done more than forty CTA interviews, so the data set was large and varied. Our project leader was Rebecca Pliske. She came to each analysis session with a photocopy of the client's goals: four bullets on a single piece of paper. She would read the goals at each session and put that paper down in the middle of the table. And then we would set to work making sense of these very complex issues. Being reminded of what the client needed us to do helped keep us focused and on track.

Develop Analyst Skills

A third factor that can greatly enhance the quality of the analysis is the skill and experience that the CTA researcher brings to the task. In the same sense that a skilled knowledge elicitor is likely to obtain better CTA data, a skilled analyst of CTA data is likely to extract core findings from a data set more effectively and more efficiently. The skills that we have found most important fall into three categories:

1. The CTA researcher has experience working with qualitative data and is knowledgeable about both qualitative and quantitative methods of analysis. An experienced CTA researcher will be able to see the possibilities in the data set and have requisite skills for identifying, documenting, and representing key insights.
2. The CTA researcher is knowledgeable about relevant findings on cognition and understands the conceptual basis for the project. The analysis team cannot recognize mental model formation, story building, sensemaking, or a host of other cognitive content in the data record if they aren't familiar with the constructs.
3. The analyst has an appreciation of the real-world context—the operational environment—that participants are describing in various ways in the data record.

    We hope that this book helps people acquire these very skills.

Summary

The intent in this chapter has been to open the "black box" of analysis and representation, and examine the processes and procedures that can be used to transform specific data elements to general findings. Reading about CTA is likely to lead to an impression that knowledge elicitation activities constitute the entire CTA process and that once the data have been gathered one can move easily, with little time or effort, into application of the findings. For a number of CTA techniques, the knowledge elicitation methods do produce a data structure and populate it. In some methods, such as concept mapping or the simulation interview in ACTA, knowledge elicitation and knowledge representation are inherently linked. But these techniques still produce representations on an individual level, specific to a particular SME, simulation, or event. They do not circumvent the necessary steps of integrating and synthesizing data, extracting meaning, and identifying key findings. *How* those activities are carried out, and how findings and applications link back to individual data elements, requires consistent attention and clear explanation.

II   Finding Cognition

The study of cognition—the way people think—has been carried out in the laboratory and in field settings, through basic research into the mechanisms of thought, and through applied research on a wide range of problems. The techniques that have been used to carry out this work include Cognitive Task Analysis (CTA) and many other methods and tools as well. You may find yourself wondering what relevance this work has for you, as a CTA practitioner. The issue is an important one for understanding the cognitive target: what are we after when we try to capture cognition?

The better and more thoroughly people understand the cognitive aspects of performance, the more successful they will be at using CTA tools to learn how people think about the tasks they perform. We have to understand the quarry in order to design appropriate methods for its capture. If we want to set a trap, we have to determine what we are trapping: eagles that need to be tagged, coyotes in the yard, grizzly bears that have wandered past their park boundaries, or salmon in a stream. We don't just set traps and hope that something interesting will wander into them.

The purpose of this chapter is to describe various aspects of cognition. In it, we explore what it means to study cognition in a natural context and how that differs from laboratory-based research. Because many CTA studies are designed to discover what experts know, we describe the nature of expertise and its cognitive elements.

We introduce a level of cognitive phenomena—referred to as *macrocognition*—that emerges when we shift the focus to natural contexts. These are the types of cognition that CTA methods are uniquely designed to capture. We provide some background about how researchers came to be interested in macrocognitive topics such as naturalistic decision making, sensemaking, problem detection, and replanning. We provide a perspective on how use of CTA methods to explore these types of cognition meets the criteria for scientific inquiry.

Cognition in Context

To understand cognitive functions, we have to appreciate the context in which they are carried out. Returning to the metaphor of trapping an elusive quarry, we have to do more than identify the quarry—we have to study the landscape in which it roams. If we are going after a grizzly bear, we need to map its territory—where it fishes, where it sleeps, where it forages—in order to decide how to set our traps.

As shown in table 8.1, cognition doesn't happen in a vacuum.[1] People's thinking is goal-directed—and usually reflects several goals at once. Even in a very trivial task, such as trying to balance the family checkbook, people may be motivated by several

Table 8.1
Features of the cognitive landscape

The *purpose* of the cognition	The reason we are engaged in thinking. Typically, people have several goals, and they may have to prioritize or trade these off against each other. Goals are often ill defined, so there is also the cognitive task of defining and clarifying goals while working to reach them.
The way we use our *prior experience*	The way a novice tries to think through a problem can be very different from the reasoning shown by an experienced performer.
Features of the *situation* itself	These may include the amount of uncertainty about key data elements, the degree to which events are changing, time pressure, the importance of the consequences, and/or whether the work involves a bounded system such as a nuclear power plant or a relatively unbounded situation such as conducting peacekeeping operations in a foreign country.
The nature of the *challenge* we are facing	Is the task a problem that is well defined, such as a mathematics puzzle, or ill defined, such as writing a song?
The *tools* that are available	These can include a calculator for helping with the checkbook or a pen for writing in the margins of a book. Tools can also be adapted from one purpose to another. For example, commercial airline pilots have been known to use an empty Styrofoam coffee cup as a memory aid (they place it over a handle for lowering their flaps as a reminder to complete other operations first).
Team members we share tasks with	These can be work partners who may be helpful or distracting.
Organizational constraints	These constraints can include procedures to be followed, what counts as an acceptable rationale, and what evidence is needed in order to raise an objection.

different simultaneously active goals. We want to get the checkbook to balance, figure out where the error had crept in (hoping that it was our spouse, not us, who had generated the error), and finish the job quickly even if it means living with a minor discrepancy of less than a dollar. The goals of wanting an accurate calculation and wanting an explanation for the error can conflict with a goal to finish quickly in order to get to sleep. Part of the cognitive landscape is the way in which people resolve these goal conflicts. Vicente (1999) has discussed the importance of describing the context of cognitive work, particularly the network of relevant goals.

Thinking also reflects our past experience. In trying to get the checkbook to balance, we may remember writing most of the checks. If we recall being rushed or confused during one of the exchanges, we may examine it more carefully. We may recall times when we hit the addition (+) key instead of the subtraction (−) key on the calculator by mistake, leading to a search for entries that are exactly half the discrepancy. Or we may remember times when the prior month's balance was the source of an error, prompting a review of the earlier arithmetic. Sometimes there are nonobvious bank charges, such as a fee for a safe deposit box—we need to look at the statement more carefully. Sometimes we have been guilty of writing a check and failing to enter it—so we have to confirm that all the checks are accounted for. Our repertoire for error checking has become far more sophisticated than when we opened our first checking account.

The remaining entries in table 8.1 are also important in trying to appreciate why people make inferences and decisions the way they do. The challenges they face, any unusual or influential features of the situation, the tools at hand to help with memory or other functions, the makeup of the team with which they are coordinating, and constraints imposed by the organization—all of these can affect the cognitive phenomena we are trying to understand.

Controlled laboratory studies can be useful for understanding individual elements of the cognitive landscape and for identifying component or underlying processes. To understand the elements as they naturally occur and interact requires field studies. The features in table 8.1 can be difficult to re-create in laboratory settings; but all of them can have an important influence on the way people think. Field studies offer a means to investigate cognition in natural settings and to grapple with the variety and complexity that is inherent in real-world circumstances.

One reason that field studies are so informative is that they allow us to examine highly skilled performance in the context of real-world tasks and settings. Expertise is a significant factor for determining what features of the cognitive landscape look like and how they play out. Learning to recognize cognition in real-world tasks often means understanding expertise and how experts and novices differ.

Expertise and Cognitive Task Analysis

Most CTA studies are conducted with subject-matter experts (SMEs) such as skilled fire-ground commanders, seasoned military leaders, experienced nurses, and highly trained pilots. These domain specialists have knowledge and understanding that distinguishes them from peers and coworkers. The CTA study is designed to elicit the knowledge and wisdom they have acquired. In conducting CTA interviews, appreciation for the nature of expertise will alert the interviewers to comments that need to be probed in more detail.

What is the basis of expertise? Do experts simply know more facts and rules than the rest of us? That is part of the story. Domain experts may have ten to twenty years of experience in which to learn facts, relationships, mechanisms, and routines, along with the contextual sensitivity to appreciate how to apply this knowledge and how to adapt it. But there is more to expertise than time spent on the job. Simple accumulation of practice is not sufficient. In research on firefighters (Klein, Calderwood, and Clinton-Cirocco 1986), we observed that ten years with a rural volunteer fire department were not as valuable for skill development as a year or two in a decaying inner city. Urban firefighters are exposed to a wider variety of fires and a vastly higher incidence rate than rural firefighters are. Although some minimum amount of time is necessary, it must be accompanied by a chance to accumulate a varied set of experiences. Expertise also requires that the person pay attention to the experiences they are accumulating and actively work toward developing and honing skill. True experts, people at the very highest levels of capability, develop their skill through actively engaging the environment, testing themselves and their performance again and again, and endlessly practicing requisite skills. Years of experience matter, but so does deliberate practice (Ericsson, Krampe, and Tesch-Römer 1993).

When a person attains a high level of proficiency, we expect to see certain characteristics of performance. We expect the person to be able to make judgments and discriminations that are difficult for most other people. The expert must be able to apply the experience to a wide range of tasks encountered in the domain, including nonroutine cases that would stymie people who are merely competent. The best experts will set the standards of ideal performance for a domain.

Finally, experts don't simply know more. They know differently. The breadth and depth of their knowledge allows them to "see the invisible" and to perceive what is missing in a situation, along with what is present (Klein and Hoffman 1993). Glaser (1976a) has characterized the shift toward greater levels of expertise as follows:

- Variable, awkward performance becomes consistent, accurate, complete, and efficient.
- Individual acts and judgments are integrated into overall strategies.
- With perceptual learning, a focus on isolated variables shifts to perception of complex patterns.
- There is increased self-reliance and ability to form new strategies as needed.

Klein and Militello (2004), based on reviews of the literature on expertise, have identified a set of key cognitive elements that distinguish experts from novices.

Mental models Experts have richer mental models than novices or even proficient performers—they understand a wider range of causal connections that govern how things work and can apply them fluidly and flexibly as events change.

Perceptual skills Experts have developed perceptual skills that enable them to notice subtle cues and patterns and to make fine discriminations that may be invisible to others.

Sense of typicality Experts have accumulated patterns and experiences into prototypes so that they can judge when they are dealing with a typical event and when they are facing something that isn't quite right and needs attention.

Routines Experts have learned a varied set of routines, so they can usually find some way of approaching problems. Usually, SMEs can just plug a well-learned routine into action, but sometimes the SMEs need to alter a routine or cobble together parts of several routines. In each case, SMEs can use their broad repertoire of routines to adapt to problems.

Declarative knowledge Experts have a lot of declarative knowledge—lots of factual information, rules, and procedures that they can draw on. Cognitive Task Analysis studies generally do not compile the declarative knowledge. They can be derived more easily from manuals and textbooks.

These types of knowledge distinguish experts from the rest of us. In doing CTA research, these are some of the places to look to probe deeper into the kinds of expertise that make a difference in a domain. Experts draw on these types of knowledge to react to challenges more effectively than novices. Experts can use this knowledge to make rapid decisions, to make sense of situations (diagnosing previous events, forming expectancies, projecting future events), to plan, to rapidly generate workarounds when they need to replan, and to coordinate their activities effectively. They are adept at a wide range of cognitive processes and functions. To grasp the full extent of their capabilities requires a wider view of cognition, a view we have termed "macrocognition."

The Nature of Macrocognition

We use the term *macrocognition* to refer to the collection of cognitive processes and functions that characterize how people think in natural settings.

The term *macro*cognition stands in contrast to *micro*cognition. *Micro*cognitive processes, such as whether attention is serial or parallel, how people solve puzzles, or the errors participants (or more typically "subjects") make when they interpret syllogisms, are typically studied using carefully controlled methods and procedures. Microcognition research investigates "basic" and universal features of the way people think. These phenomena are best studied under laboratory conditions, not in the field.

When we do examine cognition in the field, we find that people are making decisions very differently from the way they do in the laboratory. They size up situations differently. They are engaged in functions and processes that we don't see very often in the laboratory. Moreover, these functions don't even seem relevant in the laboratory. We don't expect that a subject solving a puzzle will show problem detection. The experimenters have given the problem to the subject, and no number of trials will lead the subject to discover his or her own problem. The function of problem detection emerges when we look at what people actually do at work or in everyday activities.

The functions and processes shown in figure 8.1 are the types of cognitive activities that we commonly encounter when we do field research (Klein, Ross et al. 2003). These are the aspects of cognition that CTA is designed to capture. The more we understand about these aspects, the better we can conduct the CTA—the more sensitive we can be to different forms of macrocognition, and the deeper we can probe.

The framework depicted in figure 8.1 is a work in progress. Additional macrocognitive functions will be added to the set in the future. That is why we cannot limit our investigations to this particular set of functions. Instead, the goal in conducting CTA studies is to comprehend more fully the point of view of the person performing the cognitive work.

Figure 8.1 shows six primary macrocognitive functions that field researchers commonly encounter. It is important that we consider the functions at a variety of levels. For instance, we know that individuals make decisions but so do teams. We need to gain insights into both levels and into the barriers to effective decision making at both levels. Moreover, decision making often depends on artifacts such as databases and other forms of decision support systems, and we need to consider these as well.

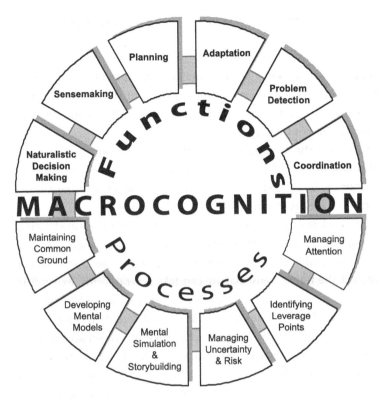

Figure 8.1
The macrocognition diagram.

Macrocognitive Functions

Naturalistic Decision Making One of the key cognitive activities in most settings is making decisions. How do people accomplish this? Field research in decision making has shown that people rarely compare options in the way that is classically prescribed. Klein (1998, 2004) has summarized the evidence taken from studies of firefighters and military commanders, showing that people typically rely on their experience to identify a plausible course of action. They then use mental simulation to evaluate that course of action without having to compare it to others. Klein's recognition-primed decision (RPD) model describes approximately 90 percent of the challenging decisions (and probably much more of the routine decisions) made in natural settings, a finding that has been replicated numerous times. (For a review of various models that have been developed within the naturalistic decision making framework, see Lipshitz 1993; Lipshitz et al. 2001.)

Sensemaking and Situation Assessment   Most natural settings demand active sensemaking. Sensemaking allows people to diagnose how the current state of affairs came about and to anticipate how it will develop in the future through the deliberate, conscious process of fitting data into a frame.

Weick (1995) has pioneered the investigation of sensemaking in organizations as a response to events that deviate from the conventional understanding of the situation. Endsley (e.g., 1995c; Endsley, Bolte, and Jones 2003) has performed numerous studies of situation assessment[2] as a means of achieving situation awareness.

Klein and his colleagues (Klein et al. in press; Sieck et al. 2004) have described a data/frame model of sensemaking that asserts that people need some sort of frame, such as a script or scenario, to understand and use data elements, and simultaneously need the data to select or construct the frame. Sensemaking can take several forms: expanding an existing account of a situation, questioning whether an existing account is accurate, explaining away inconsistent data, contrasting the merits of different accounts of the same data, replacing one account with another, and constructing a novel account.

The very large streams of data and message traffic generated by information technology result in a greater requirement for sensemaking than before. For example, Heuer (1999) studied intelligence analysts and concluded that their performance would gain more from support in making sense of messages than from gathering additional messages.

Planning   Planning is the process of modifying action to transform a current state into a desired future state. Planning functions have become increasingly complex as new information technologies provide access to massive quantities of data. Army and Marine Corps command staffs avow an organizational commitment to rational, deliberate planning processes in which multiple courses of action are generated and evaluated. However, the actual planning process seems to rely on a recognitional approach to planning, relying on the commander's expertise (Ross, Klein et al. 2003; Schmitt and Klein 1999). Mumford, Schultz, and Van Doorn (2001) have provided a good review of the planning research literature; Thunholm (2003) has documented the benefits of recognitional planning.

Adaptation/Replanning   Organizations devote a lot of attention and resources to planning, but they spend even more of their time replanning—that is, in modifying, adjusting, or replacing a plan that is already being implemented. The critical role of replanning is demonstrated in a study by Orasanu et al. (1998) in the field of aviation. Their findings suggest that the majority of commercial aviation accidents were cases where a compromised plan was not altered in time.

Klein, Wiggins, and Lewis (2003) studied replanning in Army and Air Force command posts. They suggest that replanning is more difficult than planning because the time pressure is usually greater. Also, critical resources have already been put into play, making it difficult to reassign them in the face of changing needs. Replanning depends heavily on problem detection—noticing in time that a plan is falling apart. One of the defining features of replanning is that it requires goal negotiation: the initiating goals cannot be achieved as stated; therefore, some goals will have to be dropped, others deemphasized, and other goals substituted. Finally, replanning reduces predictability, and predictability is needed for coordination.

Problem Detection   The ability to spot potential problems at an early stage is critical in most field settings. Skilled decision makers can recognize anomalies while there is still time to avoid or deflect the consequences. How is this accomplished? Klein, Pliske, et al. (2005) compiled an inventory of incidents illustrating problem detection. The incidents include reports by firefighters, critical care nurses, Navy command and control staff, surgeons, and weather forecasters. Sometimes the accumulation of evidence passes a critical threshold and signals an alarm; however, more often early indications serve to heighten the person's awareness, so that he or she monitors events more closely and notices critical cues more quickly. At other times, contrary evidence is received and explained away until the person receives a data element that cannot be discarded. In many cases, the ability to spot a problem depends on a simultaneous reframing of the situation—problem detection is triggered by a cue to revise one's beliefs about the situation and, at the same time, a revision in beliefs allows fuller appreciation of the cue.

Coordination   We know how important coordination is for teamwork, but what exactly counts as coordination? Klein (2001) described coordination as the way the team members orchestrate the sequencing of their actions to perform a task.

Teams gain influence by drawing on a wider range of resources, knowledge, and geographical coverage. However, the advantages of multiple members has to be balanced with the coordination costs incurred—the time and effort that must be expended to control the coordination, to wait for prior steps to be completed before subsequent tasks can be initiated, and to take corrective action and redirect portions of the team who are working at cross purposes.

Klinger and Klein (1999) described a case study of a nuclear power plant whose emergency operations staff was too large. Overstaffing was causing performance decrements and higher workload. The staff kept adding new members to offset the workload problem, not realizing that the new additions were actually adding to the problem. When staffing levels were reduced, performance improved and workload went down. The coordination costs of the additional team members outweighed their contributions.

In addition to these key functions, figure 8.1 also shows a set of supporting processes that are types of macrocognition.

Macrocognitive Processes

Maintaining Common Ground   This is the continuous maintenance and repair of calibrated understanding among members of a team. It is necessary for coordination; otherwise, the team members can misinterpret intentions and messages. The concept of common ground has also been described by Clark and Brennan (1991) as a basis for communication. In natural settings, common ground is necessary for effective teamwork. Endsley (1995b) has studied a related process, the development of shared situation awareness. Klein, Feltovich, et al. (2005) describe how common ground affects coordination, and how common ground can break down.

Developing Mental Models   Although the core notion has roots that span prescientific (or philosophical) psychology and psychology of the 1800s, the notion of a "mental model" has been controversial in modern psychology ever since its introduction (Anderson 1981; Gentner and Stevens 1983). We suspect this is because cognitive psychology is concerned primarily with microcognitive research, aimed at revealing cause-effect relations among fundamental (or what are believed to be fundamental) cognitive operations, such as attentional shifts and access to items stored in long-term memory. A mental model is a macrocognitive phenomenon of conscious experience, having aspects of mental imagery and aspects of event comprehension. Events in the mental model/image are formed on the basis of abstract knowledge of domain concepts and principles. So, mental models are also akin to the psychological notion of memory "schemata." However, mental models are not stored like templates, ready to be taken off the mental shelf and used in isolated acts of recognition; they are actively, deliberately formed anew each time a data set or situation is perceived. They are how sense is made of situations (Klein, Phillips, et al. in press). As people gain experience, they are able to build richer, more accurate, more consistent, more coherent mental models. We see evidence that decision makers form mental models in nearly all domains in which CTA has been applied. Weather forecasters, for example, form rich mental models, involving air masses, fronts, and the like in a four-dimensional imagining. Indeed, meteorologists themselves discovered a need to refer to a notion of mental models, to distinguish how forecasters understand weather from the computer models that provide them guidance in composing forecasts (see Hoffman, Trafton, and Roebber 2006).

Mental Simulation and Storybuilding   Mental models can be used to mentally project into the future. We tell ourselves stories about the situation as it unfolds, and we mentally explore alternative, hypothetical futures. Whereas mental models provide a causal understanding of how situations came about and what they are in the present, mental simulation involves enacting series of events and pondering them as they lead to possible futures. Like mental models, mental simulation and storybuilding are essential to sensemaking, problem detection, and decision making.

Uncertainty and Risk Management   Uncertainty is a state, and a feeling, in which we do not know or understand something, but feel that we need to, or that we should. It is a state we are in when critical data are missing, or when our goals are unclear, or when problems themselves are not clearly stated, or when we are not sure what to do next. Managing uncertainty is essential for working in ill-structured and ill-defined domains. Schmitt and Klein (1996) identified different aspects of uncertainty: uncertainty resulting from missing data, from data whose validity is unclear, from ambiguity over competing situation assessments, and from complexity that interferes with sensemaking. People may try to reduce uncertainty, but in many cases it takes too long to gather all the data. Therefore, decision makers have to act in the face of uncertainty, and have to develop skills for coping with uncertainty.

Identifying Leverage Points   Identifying leverage points is the ability to identify opportunities and turn them into courses of action. Klein and Wolf (1992) studied firefighters and Navy commanders and observed that they were not generating options by searching through a predetermined problem space. Instead, they were using their experience to identify promising leverage points in a situation and looking for ways to take advantage of these leverage points to construct a plan of action. Option generation relies on experience to construct a course of action as well as memory search.

Managing Attention   Attention management is the use of perceptual filters to determine the information a person will seek and notice. Attention management is particularly important as information technology dramatically increases the flow of data and people continually have to ward off interruptions and make decisions about how to allocate their attention.

The macrocognitive functions and processes shown in figure 8.1 are a frame for CTA studies. We take it as a given that people, whether they are pilots, nurses, military commanders, or teams of firefighters, have to perform these functions—make decisions, anticipate problems, make sense of situations, revise plans, coordinate. They use mental models, manage attention, repair common ground, spot leverage points. The job of CTA is to discover how all of this happens. For example, the CTA study cannot end with the statement that intelligence analysts have to engage in sensemaking. That is the beginning of the inquiry, not the output. The question would be how intelligence analysts make sense of different data elements, how they use, interpret, and ignore data, how they seek information, and how skilled analysts can make inferences that other analysts would miss.

Cognitive Flow

To this point, we have presented macrocognition as a set of discrete functions and processes. We have indicated that the macrocognitive elements are sometimes linked. Presented as a list they appear to operate as separate, distinct elements. In reality, people don't perform only sensemaking or coordination; they don't plan or make decisions. They do all these things, sometimes simultaneously, sometimes in overlapping segments, sometimes favoring one element over others in order to expend less energy on those that seem less important. The term *cognitive flow* describes the way a person's cognitive functions and processes overlap and change over time, a phenomena first described by William James (James 1890). The macrocognitive functions are not discrete phenomena, and they don't occur in orderly sequences, although one might get that impression from many studies of cognition (Ross and Shafer 2004). Much of that research, whether carried out in controlled laboratory experiments or conducted as field research, focuses on a few cognitive elements, and sometimes only one. Why is that?

One reason for the narrowed focus is to concentrate on a particular aspect of cognition in order to examine it, describe it, and understand it deeply. Klein's work on decision making (Klein 1989b), Endsley's extensive work on situation assessment (Endsley 1988a, 1995a, 1995b, 1997), and Weick's focus on sensemaking (Weick

1995) are examples of programs of research that have provided important insights on a specific cognitive function.

Another reason for narrowing the focus to one, or a few, cognitive aspects is the complexity of conducting CTA on the full array of cognitive processes and functions simultaneously. Researchers manage that complexity by winnowing the cognitive flow into a manageable subset of macrocognitive functions and processes. The inclination to treat and interpret complex circumstances and topics as simpler than they really are has been described by Feltovich et al. (2004) as the "reductive tendency." Dimensions that are likely to induce the reductive tendency include processes that are: continuous as opposed to discrete; dynamic as opposed to static; simultaneous as opposed to sequential; interactive as opposed to separable; nonlinear as opposed to linear; multiple as opposed to singular. Macrocognition in real-world tasks is all of these: continuous, dynamic, simultaneous, interactive, and nonlinear.

The reductive tendency can lead us to simplify cognitive flow into the separate macrocognitive functions and processes. This can be valuable; at times it may be the only way to begin to understand a particular problem. But it's important to acknowledge that we are frequently looking at cognition through a soda straw.

We have identified two general types of cognitive flow: integrated flow and segmented flow. Integrated cognitive flow is like a symphony, with different cognitive processes and functions as the separate instruments; many different cognitive elements are engaged in order to serve a single, unified purpose. For example, a military commander directing a mission may be performing sensemaking, problem solving, using mental models, and maintaining common ground across his command staff all at the same time. Segmented flow is what we commonly call multitasking, like when we juggle the vying cognitive demands of multiple tasks simultaneously. For example, an emergency department nurse attends to the needs and problems of many different patients with different requirements at any given time.

Data collected during a training exercise at the Brigade Command Battle Laboratory, Fort Leavenworth, Kansas, illustrates the density and complexity of macrocognition (Ross and Shafer 2004). Observational and interview data were collected over the course of a multiday exercise from various members of the brigade command staff and the brigade commander. The data in figure 8.2 are based on an interview conducted with the brigade commander on day two of a four-day exercise. The commander's expectation for what had occurred overnight did not match the operational picture he encountered as day two of the exercise began.

The particular event depicted in figure 8.2 covered approximately thirty minutes at the beginning of an exercise that took place on Monday. On the Saturday before, the

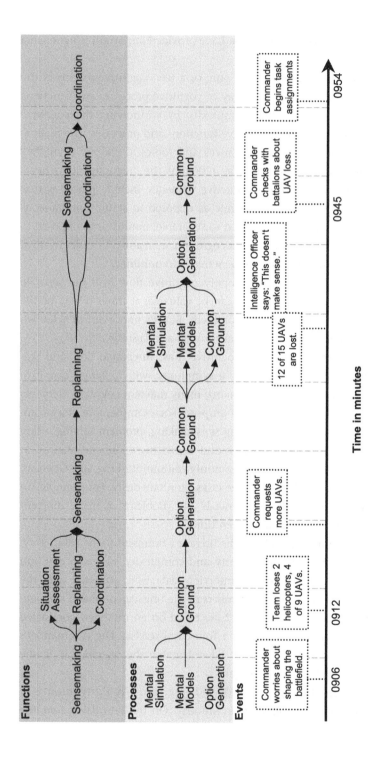

Figure 8.2

Integrated cognitive flow during a training exercise.

team participating in the exercise had planned the Monday battle and had configured the simulation to run over the weekend. However, there was a problem with the simulation, and the Monday exercise began with conditions nobody expected. Thus, the run provided a good example of dynamic replanning and also contained instances of other macrocognitive tasks.

At the end of the day's exercise, interviewers used a number of interview methods to understand what the brigade commander was thinking and doing. The primary method was an adaptation of the CDM designed to elicit information about naturally occurring macrocognitive processes and map them over an event (Ross, McHugh et al. 2003). The brigade commander reported his view entering the battle run that morning: "My thinking was we have a final little bit of preparation in the shaping phase to be able to execute the decisive action pretty quickly. It didn't take long at all to figure out that was not going to be the case."

Figure 8.2 illustrates a macrocognitive view of the commander in action during the initial period of the battle run. It shows a variety of cognitive functions and processes emerging and concluding as he struggles to make sense of the unexpected situation he encounters, receives information about additional losses, discards his mental model of how the battlefield had been shaped, replaces it with one that matches the current situational picture, and disseminates critical information to his staff.

The project was an initial effort to expand the notion of macrocognition to reflect how multiple cognitive events occur together in the real world. The nature of complex cognition is that it occurs in packages of functions and processes rather than single, sequential entities or causal chains of such things as long-term memory. The cognitive elements emerge in a fluid and flexible manner and shift readily in response to the dynamic nature of the environment. Using CTA tools to capture and represent cognitive flow is essential to having a true picture of a mind at work.

The Scientific Status of Macrocognition

Can field studies of cognition count as science? Researchers aren't necessarily testing hypotheses or contrasting experimental groups with control groups, and people are called "participants" rather than "subjects." The data can be qualitative rather than solely quantitative, and inferential statistics are rarely needed. Just about everything we associate with scientific activities in laboratory psychology is missing.

Yet these naturalistic studies do seem to share the same scientific values found in controlled experimentation. These values include objectivity, replicability, falsifiability, generalizability, observability, parsimony, skepticism, and the goal of finding causal explanations. Regarding *objectivity*, field researchers adopt protocols so that their data collection activities can be taken into account in interpreting their observations and so that their subjective impressions can be separated from objective features of the situation. *Replications* are important in field research—the many replications of the RPD findings (see Klein 1998) were critical in the acceptance of that model of decision making. *Falsifiability* is also important. One of the reasons the RPD model has gained such wide acceptance is that it allows for testable hypotheses. For example, research on the RPD model has tested assumptions about whether the model holds under conditions quite different from those that describe the original studies, such as large-scale events, events without time pressure, and decision making by novices.[3]

Generalizability is, if anything, more strongly valued in field research, particularly because this research is usually sponsored by organizations that wish to apply the findings as broadly as possible. *Observability* is pursued in the way notes are taken, in tape recording interviews, and even in video recordings of actual incidents (e.g., Omodei, Wearing, and McLennan 1998). *Parsimony* is manifested in pressure to present the most straightforward models needed to account for observations of both behavioral and cognitive phenomena. *Skepticism* may be a general value of scientists, but in practice it is best fostered by having competing theories and viewpoints oppose each other in interpreting the same data. As models of macrocognitive phenomena proliferate, skepticism and debate should increase. Finally, field researchers are keenly interested in *finding causal explanations* for their data. In short, field researchers sign up to all the scientific values adopted by laboratory experimenters. If there is a difference in the nature of the science, it is not about the scientific values.

Those who advocate the study of cognition in the academic laboratory and those who advocate the study of cognition in natural contexts differ in their views about which aspects of empirical inquiry are necessary for an investigation to qualify as science. To approach this issue, we can rely on the standard listing of the so-called stages of "the" scientific method (described in Hogarth 2001):

1. Selection of a question or phenomenon to study,
2. Observation,
3. Formulation of hypotheses,
4. Testing hypotheses, and
5. Generalizing to broader theories or other applications.

The critical stage is number 4. We cannot here go into the details of philosophy of science on this point, but it is generally held that the only method humans have ever created to allow the refutation (or disconfirmation) of hypotheses is the experiment.

For a procedure to qualify as an experiment, variables that are believed to be metrics of effects must be operationally defined and measured, and variables that are believed to act as causes must be selected or manipulated. Other variables that likely play a causal role, but are not of immediate interest, must be controlled (e.g., eliminated or held constant). Thus, manipulation and control are critical.

Here we find an apparent disconnect between field studies and laboratory studies. In order to accomplish the first two stages, macrocognition researchers primarily work as explorers, searching for interesting phenomena and investigating their nature. In contrast, laboratory scientists focus on stage 4, tinkering with nature to see how she behaves.

From our vantage point as field researchers (who have also conducted experiments when appropriate), it strikes us that testing theories by setting up controlled investigations is an important aspect of science, but is not sufficient. If it were true that an investigation could not qualify as science unless it involved control and manipulation of variables, then astronomy, for instance, would not qualify as science. Astronomers cannot manipulate the mass of stars to test a theory of their internal structure. But astronomers can disconfirm hypotheses. For instance, they can locate a great many stars, select only those with certain distances or masses or other important characteristics, and then look at their spectra to see if an hypothesis about their structure holds up (e.g., their output of X-rays should be within a certain range if hypothesis X is true.).

In his day, Charles Darwin was referred to as a "natural philosopher," primarily because the sciences were still splitting off from philosophy. Naturalism was an approach that favored empiricism, holding that knowledge comes from observation and understanding of the world, in contrast with "rationalism," which holds that knowledge comes from armchair theorizing and the dictates of authority. Even as a schoolboy, observing is something Darwin did to an extent that was unparalleled. As the voyage of the *Beagle* proceeded, a great deal of collected material and specimens were shipped back to the British Museum. Indeed, by the time the *Beagle* returned, Darwin was widely regarded as the greatest living geologist. In his research, he formulated hypotheses and even made a great many "predictions." The most famous of his predictions: He asserted on the basis of fossil evidence that at one time there must have been a creature having some of the features of reptilians and some of the features of birds. It was not long after that prediction that a specimen of archaeopteryx (a birdlike creature having teeth) was identified in a museum. None of the staff had known what to do with it.

Other researchers who were inclined toward what we would call naturalism include Aristotle, Francis Bacon, and Jean Piaget. Their science involved exploring hypotheses by observation, collection, and categorization. All of them conducted experiment-like investigations. The fact that not all of their investigations would qualify as a "laboratory experiment" does not mean that what they did was not grounded in the philosophy of empiricism and was not science.

In the case of CTA, observing things in the field setting equates to conducting naturalistic studies in the ecology of human work. Like previous forms of naturalism, the focus of CTA is on collecting observations, trying to make sense of them by forming categorizations and hypotheses, and then subsequently trying to evaluate the hypotheses in further observation. Some of this can involve selecting variables (e.g., novices versus experts), some can involve manipulating variables (e.g., designing "tough case" versus "prototypical case" scenarios for studies of reasoning). But, as we said above, the stages of "the" scientific method present a misleading picture of science. We see the "naturalistic versus experimental," the "laboratory versus field," and the "applied versus basic" distinctions as being superficial either/or schemes that do not do any justice to the true complexity of the scientific process or the ways in which various methods and methodologies can contribute to our empirical exploration of the world (Hoffman and Deffenbacher 1993).

That being said, it is true that laboratory work and field work, basic and applied work, naturalistic and experimental work each have different standards for inquiry. Methods used to study cognition in context have to be well designed and implemented in order to meet the best practices of empiricism (Hoffman and Woods 2000).

Summary

In this chapter we explored what it means to study cognition in natural contexts and how that differs from the study of cognition in a laboratory. We described a cognitive landscape that includes aspects of expertise. This perspective helps to uncover subtle aspects of cognition in real-world contexts. The cognitive landscape includes macrocognition—the cognitive functions and processes that emerge when the research focus shifts to natural contexts. We discussed how field studies of cognition and appropriate use of qualitative methodologies meet the criteria for scientific inquiry. These topics matter because there is more to CTA practice than knowledge of CTA tools and techniques. The best CTA practitioners spend time learning about cognitive elements of performance in order to understand a wide range of cognitive phenomena. By understanding the way people think and reason in natural contexts, CTA practitioners are more likely to recognize important aspects of cognition when they encounter them.

9 Trends and Themes in the Development of Cognitive Task Analysis: The Rise of Modern Cognitive Psychology[1]

Cognitive Task Analysis as a set of methods (and a community of practice) began to emerge in the early 1980s. In this chapter, we describe the historical origins and more recent trends and influences that have culminated in CTA.[2] Cognitive Task Analysis has roots in a number of disciplines. More than just a historical coincidence, the simultaneous emergence of CTA in a number of areas of science and applied research is a reason for its robustness. Understanding those multiple traditions provides an important perspective on CTA as a field of practice (Woods, Tinapple, Roesler, and Feil 2002).

Many people assume that CTA emerged as a consequence of the rapid evolution of computer-based technology and the enormous changes in the workplace that have occurred in parallel. In fact, CTA is deeply rooted in the history of psychology.

Beginning with the writings of John Watson (1914), the paradigm of behaviorism rose to prominence. It dominated American academic psychology for nearly half a century, despite arguments from within that the paradigm had significant limitations (e.g., Lashley 1951). But in the 1950s, the Carnegie Institute supported interdisciplinary meetings of linguists, psychologists, and computer scientists (Carroll 1953; Cofer 1979; Osgood and Sebeok 1954) that marked the "psycholinguistic revolution." Among the champions of that revolution were Noam Chomsky, who presented notions of "generative transformational grammar" and criticisms of the behaviorist approach to language (Chomsky 1959).

Notions from World War II–era applied research also had an impact on psychology. Developments in audio-recording and signal-processing technologies led to significant advances in research on speech perception (e.g., Liberman, Delatre, and Cooper 1952). Information theory and signal detection theory (e.g., Shannon 1948) led to research on perception, attention, and vigilance.

At a 1956 Dartmouth conference, a group of mathematicians and logicians charted a course for the field of artificial intelligence (AI), building on seminal work by John von Neumann (1958), Norbert Weiner (1948), and Alan Turing (1936). At that meeting,

Alan Newell and Herbert Simon discussed the novel idea of programming languages (Newell, Shaw, and Simon 1958). Marvin Minsky (1963) laid out the goals and central questions of AI. Throughout the 1960s, systems were created for solving puzzles, playing chess, and proving theorems. Edward Feigenbaum and Bruce Buchanan began to build inference engines based on procedural rules, eventually spawning the field of "expert systems" (Buchanan, Feigenbaum, and Lederberg 1971). This work contributed to the interest in expertise that was emerging in experimental psychology and educational psychology.

The computer and the associated notions of memory and programming provided cognitive psychology with its core metaphors. Computer science held promise for creating devices that could instantiate the theories of the nineteenth-century associationists going well beyond the telegraph and switchboard metaphors that had been prominent in previous decades (Hoffman, Cochran, and Nead 1990). Although many researchers in the 1950s and even up through the 1970s relied on notions of stimulus-response association, flowchart models came to be used to postulate stages of mental operations and decision making ("levels of processing") (e.g., Atkinson and Shiffrin 1968; Waugh and Norman 1965).

These trends converged at a 1956 conference at MIT. Newell and Simon presented their theory of symbolic information processing; Noam Chomsky laid out his theories of language; and George Miller presented his classic research on short-term memory limitations. Miller regards this meeting as marking the beginning of cognitive psychology as a discipline (Baars 1986).

A 1960 conference supported by the Social Science Research Council brought Jean Piaget's work to the attention of American developmental psychologists (see Flavell 1963) at the same time that seminal research on language acquisition was being reported by Roger Brown and Jean Berko (1960). Also in 1960, George Miller and Jerome Bruner instituted the Center for Cognitive Science at Harvard, "using the word 'cognitive' defiantly" (Miller 1979:11). Miller and other psychologists had begun to study mental imagery memory and mnemonics (see Hoffman and Senter 1978; Paivio 1975), a topic that had been largely neglected since the studies by Francis Galton in the 1800s.

In 1935 T. V. Moore had published a textbook titled *Cognitive Psychology*. This text integrated much of what was known about cognition, including ideas of mental representation, research on memory, "the association of ideas," and related topics. But to many psychologists, what marked the arrival of cognitive psychology was the publication of a book by that same name in 1967 by Ulric Neisser. There followed a number of additional important works, including Anderson and Bower's (1973) *Human Associative*

Memory, Walter Kintsch's (1974) *The Representation of Meaning in Memory*, and Craik and Lockhart's (1972) publication of their "levels of processing" theory. A number of new journals appeared, including *Cognition* (in 1972) and *Cognitive Psychology* (in 1976). Few American graduate schools at the time had courses in cognitive psychology. Courses on the psychology of learning spent as much time on rat research as on human research. But Miller, Bruner, Neisser, Bower, Kintsch, and the others made it scientifically respectable to study cognition. Newell and Simon also helped to mitigate the distrust American academic researchers had in studying mental events. If the events could be simulated on a computer, and the simulated performance could be matched against parameters of human performance, then the computer program could be considered a theory of mental activities.

This academic climate profoundly shaped the development of CTA methods. However, the strategy of using CTA methods to study cognition in field settings came primarily from applied programs. In the following sections, we describe how work in a number of disparate fields of studies led to the emergence of CTA as a discrete methodology and field of practice.

Converging Trends

The emergence of the field of cognitive psychology prepared the way to explicitly study cognitive functions, but framed this research as controlled and laboratory-based. A variety of trends and problems falling at the intersection of cognition, collaboration, technology, and work became salient between 1980 and 1985. At that time, there were multiple parallel and independent origins of communities of practice where coinage of the term CTA seemed to come naturally. These trends and fields of study include:

- Cognitive systems engineering
- European work analysis
- Instructional design
- Cognitive architectures, computer simulation, and human-computer interaction
- Ethnography of workplaces and cognitive anthropology
- Cognitive machines and artificial intelligence
- Cognitive field research and naturalistic decision making

Cognitive Systems Engineering

Cognitive systems engineering emerged in response to accidents such as the Three Mile Island accident of 1979. This showed the need for academic cognitive psychologists and human factors engineers to broaden their horizons and study human cognition and performance in complex, high-consequence settings. At the same time, advances in computer graphics and computer technology provided a basis for creating new support systems and interfaces for process control (Hollnagel, Mancini, and Woods 1986). A new generation of cognitive psychologists confronted cognitive work in control centers and tried to extend concepts from psychology to deal with actual practitioners performing substantive tasks with many kinds of tools (Woods 1996, 1998).

The designation of CTA as a method became common parlance in the early 1980s, following Three Mile Island (Hollnagel and Woods 1983). In cognitive engineering, the goal of CTA has been to reveal patterns and principles in human-computer interaction, especially how the behavior of practitioners, such as controllers in nuclear power plants, is adapted to the constraints imposed by the domain, organizational goals and pressures, and characteristics of the information technology (Hollnagel and Woods 1983; Norman 1993; Sarter, Woods, and Billings 1997; Woods and Roth 1988). The CTA process was more than the use of a single technique to examine cognitive and collaborative work. Research addressed themes that cut across phenomena and particular application domains including anomaly response, automation surprises, and how to make intelligent systems team players.

The rise of cognitive engineering coincided with the emergence of work analysis in Europe (De Keyser 1992).

European Work Analysis and the Ethnography of the Workplace

Task analysis originated in European applied psychology and industrial psychology. There, the study of work from a psychological perspective was not punctuated by behaviorism, and so there was, in effect, no such thing as "behavioral task analysis" in the European view. Programs of research in France and Belgium used approaches to ask analysis that took cognition into account (e.g., Christensen-Szalanski 1993; De Keyser 1992; De Keyser, Decortis, and Van Daele 1988; Galegher, Kraut, and Egido 1990; Galegher and Kraut 1990). In Denmark, an engineer, Jens Rasmussen, and his colleagues at the RISØ National Laboratory in Denmark made some important inroads in the engineering aspects of safety in the nuclear power industry, involving observations and interviews in the workplace (e.g., analyses of prototypical problem scenarios, Rasmussen 1986; Rasmussen and Lind 1981; Rasmussen and Rouse 1981). Research was conducted on domains including aviation safety and electronics troubleshooting, and these investigations revealed additional aspects of human problem solving and strategic reasoning (Rasmussen 1992).

One of the many reasons that the work of Rasmussen and his colleagues stands as a landmark is because of their realization that the analysis of work in complex sociotechnical systems needs to examine work from a number of perspectives (see Rasmussen, Pejtersen, and Schmidt 1990; Vicente 1999). This includes the study of

- the larger organization's values and goals (how roles are allocated to individuals, how the organization is managed and coordinated),
- the work domain itself (e.g., the analysis of problem spaces), and
- the cognitive capacities of the human worker (mental models, levels of expertise).

Rasmussen's integrative perspective on cognitive work has had an impact on many researchers, including many Americans who conduct CTA. In addition, European work analysis and ethnography quickly built connections with the first generation of cognitive engineers working in the United States.

Instructional Design

Concurrently, a group of researchers affiliated with the Learning Research and Development Center at the University of Pittsburgh and the Psychology Department at Carnegie-Mellon University launched a number of research projects regarding the nature of expertise (Chi, Feltovich, and Glaser 1980; Glaser 1976b, 1984; Means and Gott 1988) and issues of instructional design in both educational contexts (e.g., elementary school–age mathematics word problems, college level physics problems) and technical contexts in military applications (e.g., problem solving by electronics technicians) (Glaser et al. 1985; Katz et al. 1998; McKeithen et al. 1981).

Methods, referred to as behavioral task analysis, that had been used in curriculum design began to seem inadequate and incomplete in that they made little or no reference to cognitive structures (Shalin, personal communication). They did not capture domain knowledge and reasoning strategies, such as the ways that some learners seem able to effectively skip parts of behavioral task sequences. Research on problem solving circa 1968–75 (e.g., the study of computer-aided instruction, Loftus and Suppes 1972) pointed to a need to study underlying processes (of reading comprehension) and the knowledge structures involved in the mathematics domain. The new research methods that were used evolved out of the decomposition of problem-solving behaviors in terms of "learning hierarchies" (Gagné 1968, 1974), that is, sequences of learning tasks arranged according to difficulty and directional transfer. This occurred just when cognitive science was on the ascent. A 1974 symposium on cognition and instruction (Klahr 1976) included investigations using methods that could be considered CTA (see chapters by Greeno, Gregg, Resnick, and Simon and Hayes in Klahr 1976). The notion of CTA was seen as a natural contrast with behavioral task analysis (Greeno 1978).

Researchers in the field of expertise studies began to use the term CTA to refer to the process of identifying the knowledge and strategies that make up expertise for a particular domain and task (Glaser et al. 1985). Study samples shifted from naive, college-age participants who participated in artificial tasks using artificial materials (in service of control and manipulation of variables) to highly skilled, domain-smart participants engaged in tasks that were more representative of the real world in which they practiced their craft (e.g., expertise in manufacturing engineering, medical diagnosis, taxicab driving, or bird watching) (Chi, Feltovich, and Glaser 1981; Chi, Glaser, and Farr 1988; Ericsson and Simon 1993; Hoffman 1992; Shanteau 1992).

Investigators began to shift their attention from cataloging biases and limitations of human reasoning in artificial and simple problems (e.g., statistical reasoning puzzles, syllogistic reasoning puzzles) to the exploration of human capabilities for making decisions, solving complex problems, and forming "mental models" (Cohen 1989; Gentner and Stevens 1983; Klahr and Kotovsky 1989; Neisser 1982; Scribner 1984; Simon 1973; Sternberg and Frensch 1991).

Cognitive Architectures, Computer Simulation, and Human-Computer Interaction

Separate, but related to cognitive engineering, were the efforts to understand human-computer interaction to support the design of interfaces. Research attempted to map mental mechanisms at a microcognitive scale onto specific tasks. This emerged following Card, Moran, and Newell's (1983) work on the keystroke model and the later GOMS (goals, operators, methods, and selection rules) model. The work on cognitive architectures allowed researchers such as David Kieras and Peter Polson to develop cognitive simulations of the microstructure of the cognitive work required to accomplish a task (e.g., Kieras 1988). Early successes predicted differential learning times for different human-computer interface designs. Later work examined how task demands could impose working memory bottlenecks that might lead to predictable errors.

Ethnography of Workplaces and Cognitive Anthropology

Ethnographic study of the workplace emerged along with cognitive anthropology and an interest in understanding how work cultures are affected by technology change. This led to field observation of practitioners at work in their world, ethnographies of work (e.g., Cross, Christiaans, and Dorst 1996; Dekker, Nyce, and Hoffman 2003; Hutchins 1980, 1990, 1995a; Jordan and Henderson 1995; Suchman 1987), and studies of "situated cognition" (Clancey 1997, 2001) and "cognition in the wild" (Hutchins 1995a). Classic studies in this area include Orr's (1985) study of how photocopier repair technicians acquire knowledge and skill by sharing their "war stories" with one another, and Lave's (1988, 1997) study of the traditional methods used in crafts such as tailoring and his study of the nature of math skills used in activities like shopping and dieting. Lucy Suchman's work (1987) is generally noted as a landmark; it resulted in an explosion of interest in ethnography of the workplace (e.g., Barley and Orr 1997). Suchman asserted that many problems are solved on the fly, utilizing resources that are inherent in the problem context. In contrast to notions that problem-solving behaviors are structured by preformulated mental representations and procedures, such as plans (c.f., Miller, Galanter, and Pribram 1960), situated cognition theorists advocate a view that such representations are "best viewed as a weak resource for what is primarily ad hoc activity" (Suchman 1987: ix). As Knorr-Cetina (1993) argued, a good understanding of a domain and the operatives who work within it is unlikely "to be gained from observation alone." She continued, "We must also listen to the talk [by operatives] about what happens, the asides and the curses, the mutterings of exasperation, the questions they ask each other, the formal discussions and lunchtime chats" (1987:21).

Related to cognitive anthropology is a field known as the "sociology of scientific knowledge" (e.g., Barnes 1974; Collins 1993, 1997; Fleck and Williams 1996; Knorr-Cetina 1981; Latour and Woolgar 1979; Lynch 1991, 1993; Williams, Faulkner, and Fleck 1998). Researchers within this approach proposed that the acquisition of scientific knowledge is as much a social accomplishment as a process of objective empiricism, and thus argued that science is a largely constructive process that cannot be analyzed without consideration of the historical, cultural, and social context in which it occurs.

Cognitive Machines and Artificial Intelligence

Computer simulation of thinking (Newell and Simon 1972) seemed to require methods of protocol analysis, but another strong motivation to develop methods was to create expert systems using knowledge elicitation methods. Researchers sought to compile the production rules elicited from subject-matter experts (Hoffman 1987) in order to model the way experts perform complex tasks. Literally hundreds of domains were the subject of expert systems development efforts. In the development of expert systems, there must be some sort of knowledge elicitation procedure as one component to the total process of knowledge acquisition (Regoczei and Hirst 1992). Knowledge-elicitation procedures, such as those used in expert system development (for reviews, see Cooke 1994; Gordon and Gill 1997; Hoffman et al. 1995; Olson and Biolsi 1991), could be considered a class, type, or example of CTA.

Cognitive Field Research and Naturalistic Decision Making

Attempts to apply theories and methods from the field of judgment and decision making to complex, real-world settings led to new methods for studying decision making and new models to describe decision making in those settings (e.g., Klein 1989a). Researchers who studied domains such as firefighting and clinical nursing began to note that observations from field studies of experts in action in complex settings were at odds with formal normative models of decision making that had come from research in the area of judgment and decision making. The inability of such models to account for, or even make meaningful contact with, results from the field studies prompted the Army Research Institute to sponsor a conference in Dayton, Ohio, in 1989. That conference helped define a new community of practice labeled naturalistic decision making (NDM) (Klein et al. 1993). Klein, Calderwood, and Clinton-Cirocco (1986) described the Critical Decision Method to identify and probe the challenging decisions made during critical incidents (Hoffman, Crandall, and Shadbolt 1998; Klein, Calderwood, and MacGregor 1989). NDM researchers quickly expanded their focus of interest beyond decision making to encompass a wider array of cognitive functions and processes, including problem detection, planning, situation awareness, and sensemaking.

Like the other trends, NDM dovetailed with work emerging from other arenas, particularly ethnography and work analysis. The analysis of proficient behavior had to extend beyond the laboratory to investigate cognitive activity in the field setting. Human factors psychologists found that they needed to focus on decision making in situations marked by time pressure, high risk, ambiguous or missing information, and conflicting goals (Lipshitz 1993; Orasanu and Connolly 1993; Woods 1993), which are the hallmarks of the NDM paradigm (see Klein, Woods, and Orasanu 1993).

Finding Common Ground

With the exception of European work analysis,[3] these trends all occurred at the boundaries of traditional disciplines. While we have discussed history in terms of historically separable trends, they are neither conceptually nor empirically distinct. Individuals working in each of the emerging communities of practice found themselves discovering and learning from findings from the others and rediscovering things that had been long known by the others. There has also been cross-fertilization in terms of methods as people have created innovative methods and brought them to bear on both old and new problems and have opportunistically learned about each other's work and the traditions from which they have come.

For example, studies from cognitive engineering found the same patterns that had been found by ethnographers and expertise researchers (see Goodson and Schmidt 1990; Suchman 1990; Woods 1993; Woods, Roth, and Bennett 1990). Cognitive engineering placed an emphasis on field work, which had European roots that had already intertwined to some extent with ethnography. But cognitive engineering added experimental values, since simulators and rapid prototyping technology allowed investigators to control the problems that practitioners faced and allowed the study of new artifacts and the manipulation of interface features to help reveal strategies and work practices. The tradition of critical incident studies in human factors (Flanagan 1954) was updated to a cognitive work context (e.g., Hoffman 1987; Hoffman, Crandall, and Shadbolt 1998; Klein 1989a) to provide another complementary approach in the bootstrapping process to understand what it means to practice in a field of activity.

Shared Goals

Each of these communities of practice represents an attempt to understand cognitive systems in context—how experts and teams of practitioners confront significant problems, aided by technological and other types of artifacts (Hoffman, Hayes et al. 2002). The fundamental reference point for all of these trends is the field or real-world setting.

Each of the trends we have described includes a general approach or set of specific methods to capture and apply concepts about cognition. The emphasis on cognition has occurred in response to the transformation of the workplace to put greater emphasis on cognitive and collaborative work.

The process of studying cognition at work quite dramatically changes our concepts about the goals of cognitive science. Rather than just seeking a broad, general theory of learning or cognition, one must also examine success as well as failure, expertise as well as error. What challenges practitioners? What makes situations hard? How do practitioners succeed despite the constraints under which they engage in cognitive work? How do they fail?

The process of studying cognition at work also dramatically changes our concepts about the boundaries of cognitive activity (e.g., Hutchins 1995a). New phenomena emerge as technology and organizational change transform work activities. Increasingly, studies of cognition at work do not see cognitive activity as being located or isolated in a single individual, but as distributed across multiple agents as part of a stream of activity (Hutchins 1995b; Klein 1998). Cognitive work is embedded in larger, professional, organizational, and institutional contexts. To understand the nature of cognitive work requires methods that illuminate both the cognitive activity and the contexts in which it occurs.

Summary

In this chapter we explored the historical roots of CTA. Academic research traditions and trends in applied research have each contributed to the development of CTA methods. Reaction against the dominance of behaviorism in North American psychology led researchers to a renewed interest in understanding cognitive processes. This was accelerated by the advent of information theory and computer metaphors of mind. In parallel, the impact of technology in the workplace prompted application-oriented researchers to explore a wider array of complex social-technological influences. The confluence of factors produced a wave of interest in cognitive work and led researchers in a number of communities of practice to discuss what have come to be called CTA methods.

Information technology (IT) has become a constant in our lives. It has become a part of how we work, how we shop, how we communicate, and how we learn about what's going on in the world. In the workplace, rapid evolution of IT has resulted in a number of transformations: changes in the roles that people take as they work with each other, changes in the way they perform their jobs, changes in what counts as good performance, and changes in the types of errors people are likely to make.

These transformations create a number of cognitive challenges. In designing and introducing new types of IT, there is a wide range of decisions to be made about how people will work with the technology—how they will coordinate with it, what roles the technology will play, and how to allocate functions between people and technology. In this chapter we illustrate some of the cognitive demands that are created by IT.

The Challenges of Sharing Cognitive Functions with Information Technology

To illustrate the challenges of sharing cognitive functions with IT, we present three examples: flight management systems in aviation, unmanned aerial vehicles (UAVs) for military data gathering, and infusion devices in medicine.

Flight Management Systems

Airplanes were originally designed such that the flight surfaces were controlled directly by the pilot, who operated mechanical levers controlling the flaps, rudder, and engine to vary airspeed, altitude, heading, pitch, and roll. Around World War II, developers introduced "autopilot" devices into the cockpit so that the airplane itself could sustain a course without requiring constant vigilance. Over time, these autopilot devices became more complex, taking over a larger share of the work of flying (Sarter and Woods 2000).

Today, pilots operating state-of-the-art flight management systems (FMSs) have to program these systems before the flight. The FMS is a flight-planning organizer, coordinating with other aircraft IT systems to collectively create the auto-flight ability of a technologically advanced aircraft. When corrections are needed, the pilots enter these into the FMS computer rather than manually changing the aircraft controls. The errors that pilots made with manual systems, such as failing to manage fuel or permitting the airplane to drift from the desired course, have been mitigated thanks to the tireless synthetic memory of the system. The advent of the FMS has made flying easier by reducing the number of tasks that need managing in a given time.

But new types of errors have emerged, such as entering the wrong data. Pilots can find it hard to detect keystroke errors that might not have an impact for many minutes, whereas stick and throttle controls provide immediate feedback. Olson and Sarter (2001) explain how pilots entering configuration data into the FMSs are actually making (reversible) commitments about the flight regimen. The pilots are often unaware of these commitments because they don't understand the logic of the systems they are using.

Flight management systems have helped to make aviation much safer than before while increasing fuel economy and reducing the number of crew members needed in the cockpit. They are a valuable use of IT. Yet the FMS also illustrates how IT can negatively transform the work people do. There are some aspects of flying, such as air speed, that pilots do not want to give up to automation. Pilots will maintain some manual control, but they must now also mentally simulate the flight further into the future in order to figure out what the FMS is doing, why it is doing that, and what it will be doing next (Wiener 1993) in order to avoid conceptual errors in addition to the still-present possibility of manual control errors.

Woods and Sarter (2000) introduced the term "mode errors" to describe the type of problems that arise when a pilot believes the system is in one mode (e.g., decelerate for landing) when it actually is in a different mode (e.g., compensate for icing). The same input into the FMS can have an effect that is dramatically different from what was intended if the system is not in the mode the pilot believes it is in. Pilots who have entered an incorrect keystroke or location identifier may believe themselves to be on course for potentially significant amounts of time. The pilot must also understand the limitations of the FMS—for example, an FMS will not allow a flight to return to an already-passed waypoint, precluding the possibility of flying in a holding pattern. The pilot must learn to anticipate these problems and be prepared to work around them in order to effectively use the FMS.

Flight management systems also illustrate the difficulty of predicting how envisioned systems will be experienced, which is the *envisioned world problem* described by Woods (1995) and Vicente (1999), among others. Information technology that is intended to improve the way people accomplish a task is almost certain to change aspects of the work being performed. Because the work itself has been altered, operators will implement the technology using strategies or work practices that were not anticipated by the designers. The problem is that designers are aiming at a moving target. The work they are designing tools to support is going to change when the technology is introduced, partly as a result of the technology implementation. The introduction of FMSs reduced some of the tedious aspects of flying and replaced them with new requirements that pilots understand the logic of the systems.

Unmanned Aerial Vehicles

The challenges generated by IT are not specific to aviation—they apply whenever innovative technologies are introduced. Military developers have designed a variety of types of unmanned aerial vehicles (UAVs), but we need only consider the most simple: a small drone airplane with a camera that can send a stream of images to a command center. These UAVs can replace forward observers flying light airplanes over enemy territory. Army commanders can now receive images of enemy positions prior to launching an attack without risking soldiers' lives. Nonetheless, the information provided by the UAV carries its own set of costs. Unmanned aerial vehicle operators face a number of cognitive challenges, some of them substantially different from those experienced by pilots. Accident and incident rates for UAVs are several times higher than those for piloted aircraft (Williams 2004). Many of the sensory and perceptual cues available to pilots are missing for UAV operators. Unmanned aerial vehicle operators do not have the same visual information, and none of the kinesthetic or tactile cues. Visual demands of the flight control tasks are high, and the information provided operators is complex (McCarley and Wickens 2004). For example, UAV operators must engage in mental transformations of spatial relations, since the orientation of the operator (to some reference heading in the world, e.g., northward) is not always the same as that of the UAV (e.g., it may not be flying northward), nor is it necessarily the same as the perspective of the camera, which may be mounted on the side of the UAV.

Unmanned aerial vehicles also create additional demands on military commanders, changing the way they deploy forces and requiring them to make decisions about when and where to deploy the UAVs, when and how to monitor the imagery, how to make sense of the imagery in conjunction with other types of reports, and what information and imagery to pass along to platoon leaders without interrupting and confusing them. Studies are underway to sort out these issues, but the answers will only emerge through years of experimental and actual use. Developers can anticipate some of the demands and team reconfigurations generated by the introduction of UAVs, but it is unlikely that anyone can predict how the best practices will evolve.

Medical Infusion Devices

Computer-based infusion devices are also a form of IT. These devices are becoming increasingly common in a variety of health care settings. They allow patients to self-administer medications to strict dosage levels that are determined by physicians and nurses. The issues that surround medical infusion devices were addressed in a project by Obradovich and Woods (1996). They studied a computer-based device for women with high-risk pregnancies. The device, a portable, battery-operated electronic infusion pump, injected Terbutaline[1] (a drug used for uterine relaxation) through a syringe into a subcutaneous site in the abdomen or thigh to maintain a steady and continuous level of medication. Dosage levels are critical for this treatment: underadministration can fail to control the onset of premature labor, whereas overadministration can result in toxicity that affects both the mother and the baby. Moreover, the dosages required to maintain a good treatment level have to be changed over time as a pregnancy advances. Given the many dosage issues involved, the computer-based devices must be individually programmed for the particular treatment plan of each patient.

The infusion device represents a very effective way of automating drug regimens that are otherwise difficult and time-consuming to administer. However, the device also creates its own cognitive demands: keeping track of system modes, interpreting alarms, and making sense of displays and of feedback on the state of the device. To simplify operation, one version of this particular infusion device had inputs that consisted of only four buttons. However, the sequence of button presses was not meaningful to patients, and they became easily confused about how to use the input feature. Because designers did not anticipate patients' confusion, the device did not alert patients to errors in the sequence of button presses—it merely defaulted to its normal operation screen. As a result, patients could commit errors but believe that the inputs had been successfully entered. Another type of mode error occurred when the patient tried to program the device in interval mode when it was actually in rate mode. The device accepted the inputs but interpreted the entry as rates, not intervals—a substitution that could have dangerous effects on the dosage regimen. Moreover, the device did not provide any feedback about its own behavior. Therefore, with many different operational modes and weak feedback about its status—about which mode it was actually in—the device was very prone to mode errors. The introduction of IT created a class of error that was not possible when medication was administered manually.

This medical infusion device illustrates how IT can impose new demands for making sense of situations. Patients need to have a certain degree of background information about how the device works—they need a reasonable mental model of the device's logic. They have to be able to detect anomalies and surprises. And they need firm attention management strategies to make sure they are looking at the right place on the display in order to notice various indicators. The device doesn't show the overall therapy plan. Therefore, patients have to build their own mental model of the plan they have just programmed and compare this model to the desired therapy plan. They must figure out how to recognize when they have incorrectly entered a therapy plan, or have modified the therapy inadvertently. And most often, they must quickly learn to do all this as novice users of an unfamiliar medical device. They do not come to the task with education and prior knowledge of issues that surround infusion devices.

The evaluation performed by Obradovich and Woods (1996) suggested that the infusion pump designers did not do a good job of taking into account the cognitive demands of the device operation. Information technology lets designers easily develop systems that appear workable—the system will get the job done as long as everything goes according to plan. But once a deviation is introduced, the devices turn out to be brittle, and they don't help the operators navigate through the system once things go wrong. The consequences of poorly designed medical devices can be profound, resulting in medical injury and death (Vicente et al. 2003). Understanding how the user of the device will understand the task and how he or she may become confused is part of the designer's task. Identifying these cognitive demands is essential to useful, safe design.

The Cognitive Demands Created by Information Technology

Information technology can place heavy cognitive demands on operators during normal conditions. The demands increase greatly when workarounds are needed, when anomalies arise, and when time is short.

We should point out that, of course, IT is not unique in creating cognitive demands on workers. All work is cognitive to some degree, requiring thinking, planning, and understanding consequences, even work that many people think of as mindless manual labor. For example, Shalin and Verdile (2003) studied utility service workers and found that even a simple type of manual labor such as digging ditches to lay cable placed multiple cognitive demands on the workers. They had to make decisions about the appropriate tools to use to complete a trench in an area with many buried cables (e.g., hand tools that take longer versus a backhoe, which is faster but less precise). They had to use mental simulation to foresee potential problems. For example, when putting a pipe into a ditch and fastening the slip coupling with large bolts, if the bolts were set too high, they would eventually become exposed after a few years due to soil erosion. They had to develop good mental models that they could use to form expectancies. For example, if they were using tools that weren't working properly, they would realize the problem and fix the problem or figure out a workaround. They had to use anticipatory thinking to plan—inexperienced workers often failed to load the tools they would need throughout the shift because they didn't understand how the work would be carried out once they arrived at the site. Scribner (1985) has described the complex strategies required to perform seemingly simple tasks, such as those used by milk deliverers to organize their schedules. Rose (2004) offers a compelling argument for the complex and highly integrated cognitive, perceptual and spatial skills required for many kinds of work—waitressing, carpentry, construction, welding, hair styling—that are not typically considered cognitively challenging.

All human work has cognitive elements. What is unique about IT is in the way it raises the cognitive ante. Howell and Cooke (1989) observed that advances in technology and machine intelligence had effectively increased, rather than lowered, cognitive demands on humans. As machines have taken on tasks that are highly procedural, predictable, and routine, what has remained for humans are the more complex aspects of work: tasks requiring judgment, assessment, diagnostic power, decision making, and the ability to plan and anticipate.

Moreover, with the widespread applications of IT throughout the work sector, the prevailing view of human performance has shifted as well. The limitations of traditional task analyses for describing key elements of performance have shown us just how sophisticated people's conceptual abilities are. Behavioral task analyses that attempt to decompose tasks into steps and order sequences are inadequate because skilled IT users are not following steps. We need to go beyond task decompositions and understand the users' points of view—how they are viewing the work, how they are interpreting the task, how they are adopting or rejecting strategies, and how they are modifying or abandoning standard procedures.

Work by Mumaw et al. (2000) with nuclear power plant operators shows the criticality of cognitive requirements. How does an operator know the current status of the plant? On the surface, it would seem that operators had to scan the data available in the control room and form an interpretation. But Mumaw et al. found that simply reading the instruments wasn't sufficient. Skilled operators use their mental models and contextual knowledge. They take into account the parts of the plant that are currently shut down for repair, or are due for repairs to correct malfunctions. Nuclear power plants are always in a state of repair and modification. Therefore, even if an operator had memorized all the diagrams of plant operations, and scanned all the displays and instruments, it wouldn't add up to a picture of the status of the plant without knowing which systems and subsystems were active, which were idled, and which had been malfunctioning. The operators had to form mental models of how ongoing modifications would affect the readings obtained from control room displays and indicators and take these modifications into account to determine if the plant was running smoothly or if it was experiencing difficulties. The wide variety of possible configurations and interactions of maintenance actions and problems makes it impossible to describe the operators' performance using behavioral task analysis methods. Behavioral task analysis methods can be extremely useful for addressing the steps needed to operate a system under standard conditions. Behavioral task analyses are not well suited for capturing cognitive requirements, such as when operators perform workarounds to overcome system limitations.

Given this adaptive balance between human work and machine function, it is critically important to understand the types of challenges IT poses for understanding and describing cognitive requirements. Here are several of these challenges.

Rapid Changes

We have to constantly learn and adapt to keep up with rapid changes in IT. The pace of innovation is higher in IT than in many other technologies. Twenty-five years ago, early adopters were mastering WordStar, the exciting new word processing system.[2] Spreadsheets had not yet arrived on the scene, DOS was cementing its position as the operating system of the day, and the Internet was a tool used by a small group of elite scientists and engineers. In 1992, the World Wide Web was established by the European High Energy Particle Physics Laboratory (CERN). Online dialup systems such as CompuServe and AOL that provided access to the Web only became available during the last decade.

Complexity of the Work

Even though utility service workers, carpenters, and waitresses are performing cognitive operations, their work does not approach the complexity of an intelligence analyst sorting through hundreds of daily messages to try to find critical information trends or an air traffic controller keeping track of dozens of aircraft simultaneously. Task complexity can take many forms, such as the number of different factors to track, their diversity, their level of interaction, the types of sensemaking required, and the effort needed to direct attention to high-payoff regions. By any of these standards, the information explosion complicates our work. One of the selling points of IT is that it makes so much information available. For system designers, this feature of IT creates challenges for understanding how people are going to perform (or redefine) tasks in the face of this avalanche of data. For people responsible for integrating IT into the work setting, there is a growing sense of alarm about information overload and the consequences it holds for efficient and effective task performance.

Difficulty of Developing Mental Models

Another challenge IT presents is that it can obstruct our view of how systems work and thereby create obstacles to development of accurate mental models of the system and how it fits within the larger work context. We may need to understand how a computer-based support system works, or how our work crew is going to complete a manual task, or how we will conduct an ambush differently by using reconnaissance information from a UAV. We may rely on cause/effect beliefs to create a story of how the work will be done. That story is our mental model.

The more rapid changes are to the technology and the work, and the more complex the work, the greater the struggle to figure out how things interact and how outputs are produced from inputs. When people's mental models are flimsier, the various functions that depend on solid mental models suffer correspondingly. For example, we can expect that operators of IT systems will struggle with workarounds if the standard procedures aren't effective. In the example of infusion devise users, Obradovich and Woods (1996) noted how hard it was for users to figure out the infusion pumps when standard procedures didn't seem to work. We can anticipate that system operators will often be surprised because they lack a solid basis for knowing what to expect from the IT systems and tools they are given to use.

Operator Skills

Criteria for selecting operators with appropriate skills are often clear for physical aspects of work. Strength, height, reach, and agility requirements can be specified and measured. For cognitive aspects of work, the requisite skills and abilities are not so easily identified. Even when the skills are known, the tools for measuring those skills and abilities may not exist. One consequence is that IT developers may lack information about the skills of intended users and may have to contend with considerable variability in the target audience of operators. Infusion pump operation is a perfect example of this. Designers didn't comprehend how first-time patients would make mistakes with the system, and so they didn't design in ways to catch likely errors.

Envisioned World Problem

Information technology frequently transforms the nature of the tasks it was designed to support. And that makes it hard to anticipate how the technology will actually be used. In many cases, the technology simply helps people do their current jobs more easily or effectively. Early cell phones were just telephones that we carried with us. But sometimes the technology changes the nature of our jobs. A physician responding to changes in a critically ill patient can now make a telephone call while driving to the hospital and get an immediate update on the patient's condition, coordinate aspects of care, and order needed diagnostic tests, and even view pictures of the patient. Walking into the unit, the physician is better prepared to manage the patient.

The envisioned world problem cannot be addressed by simply documenting all the tasks a person is going to perform. We have to push deeper and understand the cognitive challenges that the work presents. The tasks themselves may change as the IT is introduced. But the critical cognitive challenges of the work are more likely to endure, no matter how these challenges are handled: physicians diagnose, military commanders evaluate plans, firefighters size up the situation, and pilots are alerted to problems. Therefore, IT places demands on tool developers to provide flexibility so that the work can evolve in accord with the technology supports that are being made available.

The factors described here—rapid changes in technology, the complex nature of the work that technology must support, how technology can impede development of robust mental models, the difficulty of pegging whether operators have requisite skills and ability, and the ways in which technology alters the workplace in unanticipated ways—can all compromise the effectiveness of an IT application. In the section that follows, we describe how CTA can support IT developers in identifying cognitive challenges and understanding cognitive demands.

Using Cognitive Task Analysis to Design IT Applications

Cognitive Task Analysis becomes more valuable as the nature of the work becomes more conceptual than physical, when the tasks can't be boiled down to procedures, and when experts clearly outperform novices. Under these circumstances, the features of cognitive tasks become increasingly important in developing and implementing IT.

Cognitive Task Analysis is an obvious tool for identifying, documenting, and representing the cognitive features of performance so that they can be incorporated into the development and implementation of IT. On the surface, IT calls for support of information management because the technology generally increases the flow of information, and the operators are primarily involved in managing this flow. But it is a mistake to overemphasize information management. The purpose of the information, and the IT, is to produce better and faster decisions and judgments, more effective planning, enhanced sensemaking, and so forth. Information management is a means; it is not an end. If the information is well managed but does not have an impact on performance accomplishment, then the technology is without value—it's a toy, not a tool. We have to keep our perspective on the uses of the information, not the information itself. We have to understand the cognitive landscape that permits decision makers to effectively use IT.

When IT is being designed to provide various forms of decision support, the developers are likely to puzzle about how the intended operators are going to perform key cognitive activities. Cognitive Task Analysis methods provide a means for answering those questions. Let us examine the three examples posed at the beginning of the chapter, FMSs, UAVs, and medical infusion devices.

Flight management systems are intended to increase the situation awareness of the flight crew. However, the operation of the FMS depends on the cognitive tasks of managing the information entered into the FMS and alterations in those inputs. Flight management systems require pilots to manage attention by knowing when to attend closely to the system (as in transition periods from one mode to another). Flight management systems require problem detection in case the FMS is generating implausible recommendations due to input errors or mode errors or other difficulties. Flight management systems require pilots to engage in sensemaking to fit the FMS analysis with the instrument data. Pilots depend heavily on their mental models of the logic driving the FMS.

The operation of UAVs places heavy burdens on operators who have to manage all the information transmitted by the UAV. Depending on the sensors that are mounted, this information can be visual feeds from a camera, infrared images, alphanumeric messages, or other types of data. This information flow places demands on attention management—how carefully to monitor the UAV feeds versus other data sources. Unmanned aerial vehicle operators have to carefully manage their attention to anticipate the value of the UAV data feed at any particular time. Unmanned aerial vehicle operators need problem-detection skills to determine when a UAV may be malfunctioning or has become vulnerable and needs to be recovered as quickly as possible. Sensemaking is important for UAV operation in many ways, such as correlating the current position and heading of the UAV with the imagery being received. Unmanned aerial vehicle operation also requires coordination; several soldiers may have to operate a single UAV (one to fly it, one to interpret the data), and different team members may have to replan the course of a UAV and rapidly disseminate the data.

Infusion devices also depend on information management (inputting data into the system, modifying data the system already has, interpreting data from the system about its status). Problem detection is critical because system errors are so difficult to notice and can have such serious consequences. Patients have to engage in sensemaking to determine when and how to alter drug regimens. Patients need a mental model of the system logic in order to avoid and correct errors.

Researchers such as Don Norman (1986) and engineers such as Jens Rasmussen (1981) initiated the field of cognitive systems engineering in response to the challenges of IT and the demands of handling complex systems such as the control rooms of nuclear power plants. Petrochemical engineering requires a deep understanding of the chemicals involved along with the technology for their processing into commercially useful states. Structural engineering requires a deep understanding of the materials (e.g., steel, concrete, plastic) involved along with the technology for their arrangement to bear weight. Similarly, cognitive engineering requires a deep understanding of the cognition involved (e.g., the nature of decision making, problem solving, sensemaking) along with the IT for supporting and altering cognition in order to speed up cycles, increase accuracy, and reduce opportunities for catastrophic failures. The agenda of cognitive systems engineering has been to develop a knowledge base and a set of methods for accomplishing these goals.

Cognitive Task Analysis is an essential method for conducting cognitive systems engineering. Cognitive Task Analysis is *not* the same as cognitive systems engineering. There is a lot more to design than probing deeply into the way people think about a task, just as there is much more to petrochemical engineering than understanding the complex properties of polymers. Rather, CTA methods are aimed at discovering and describing the critical functions that are needed to perform complex activities. Hoffman, Klein, and Laughery (2002) reviewed the range of design methods that had been formulated to carry out cognitive systems engineering and found that some form of CTA appears in just about all of them. Cognitive Task Analysis contributes significantly to the design process by revealing what is going on inside the heads of people who will use the technology.

Summary

In designing and introducing new types of IT, there are many decisions to be made about how people will work with the technology—how they will coordinate with it, what roles the technology will play, and how to allocate functions between people and technology. In this chapter we described some of the challenges and opportunities that arise when IT is introduced to help people perform tasks and some of the cognitive demands that are created by IT.

III Putting CTA Findings to Use

11   The Role of Cognitive Requirements in System Development

Cognitive Task Analysis (CTA) methods are particularly well suited for developing information technology to support cognitive activities, especially the macrocognitive functions described in chapter 8. New information technology promises to improve decision making by speeding up the decisions and making them more accurate and more flexible. New technologies seek to strengthen situation awareness and sensemaking, to increase sensitivity to potential problems, to bolster adaptivity, and to support team coordination. But these technologies can deliver on their promises only if they are designed and engineered to support cognitive functions, which mandates that CTA practitioners learn what those functions are.

To accomplish this, the field of cognitive systems engineering (CSE) has emerged over the past few decades (Hollnagel and Woods 1983; Rasmussen, Pejtersen, and Goodstein 1994) as a merging of the capabilities of CTA researchers (including cognitive scientists and human factors psychologists) and technologists (including design engineers and computer scientists). Cognitive systems engineering blends cognitive science, human factors engineering, and systems engineering. Hoffman et al. (2002) have documented the range of CSE strategies currently in use. Almost all of these strategies depend on some form of CTA. After all, how can you do CSE if you don't study the demands of the cognitive work that you are trying to support?

Decision-Centered Design (DCD) is one of many CSE methods. Researchers have developed a variety of CSE techniques, such as Cognitive Work Analysis (Vicente 1999), Applied Cognitive Work Analysis (Elm et al. 2003), Situation Awareness-Oriented Design (Endsley, Bolte, and Jones 2003), Use-Centered Design (Flach and Dominguez 1995), Work-Oriented Design (Ehn 1988), Work-Centered Design (Eggleston, Young, and Whitaker 2000), and Cognitively Oriented Task Analysis (Shalin et al. 1997). Each of these approaches seeks to support a full range of cognitive functions. Each advocates the use of some form of CTA. Each faces the challenges of forging partnerships with design teams in order to incorporate cognitive data into the design process. At the same time, each has favored methods and a particular primary focus.

This chapter suggests some ways to use CTA in the system development process. We describe one process—DCD—in some detail in order to provide an illustration of how CTA can be incorporated into a process for designing, developing, and evaluating technologies that are intended to amplify and extend the human ability to make good decisions. It is our intent in this chapter to provide guidance and examples and to help readers find ways to improve the design of information technology rather than to advocate for a particular approach as superior to others.

Let's start with an example of how researchers have used CTA to formulate a better human-computer interface (HCI).

Design of a New Human-Computer Interface for AWACS Weapons Directors

An early example of designing for decision making was a research project carried out to test cognitive engineering principles, including the benefits of using CTA in interface design (Klein 1998; Klinger and Gomes 1993). The project team focused their work on the Airborne Warning and Control System (AWACS), a flying command post used by the U.S. Air Force. They set out to improve the design of the weapons director (WD) station in AWACS aircraft, focusing on the WDs' air defense mission.

The researchers conducted CTA interviews using the CDM with twenty-four WDs and observed many hours of training during exercises at Tinker Air Force Base. The data collection centered around critical decisions, particularly instances from the 1991 military campaign in Kuwait and Iraq (Desert Storm). The research team identified the tough decisions and probed these to understand not only what made them so hard, but also what kinds of errors WDs might have made and what kinds of cues and patterns the WDs used in making their decisions. The researchers combined the critical decision data to generate the summary in table 11.1. Table 11.1 is a form of decision requirements table (DRT). It lists the most challenging decision requirements and other cognitive demands. It also shows why these decisions and cognitive demands were difficult and what HCI recommendation was intended to address this difficulty for human-computer interaction.

Based on the difficulties and errors identified in the CTA, the research team generated a number of recommendations for an improved interface. The recommendations were evaluated on two dimensions: impact and ease of implementation. Four of the recommendations were judged to be both high-utility and easy to implement:

Table 11.1
Decision requirements table for the AWACS weapons director

Decision and Cognitive Requirements	Why Difficult	HCI Solution
Detect and track primary threats	Attention, screen clutter, memory, loss of understanding, must monitor tracks to determine history and possible hostile intent	Symbology for flagging major threats (e.g., high, fast aircraft)
Anchor sensemaking around key threats and assets	Screen clutter, most important tracks not identified, dynamics of situation become complex, loss of understanding	Symbology for flagging the threats and the key assets (e.g., tankers)
Estimate intercept geometry	Spatial proximity not sufficient	Automated nomination—a decision support system
Allocate resources—assign combat air patrol	Tradeoff of when to use resources, must maintain fighter flow (sufficient aircraft in reserve, on station, in battle)	Automated nomination
Maintain understanding of the situation	Screen clutter, operators must look away from scope to input actions, cannot differentiate geographical boundaries, communication workload	On-screen menu, symbology, boundary differentiation
Track identification	Secondary task, done as workload permits, with longer delays the difficulty increases	On-screen menu, automated nomination

• Better symbology would allow faster identification of threats and major assets (i.e., circles around threats and assets).

• Better use of color would help the WDs distinguish air and sea in order to reduce confusion and to support better situation awareness.

• Better placement of the most frequent and critical switch functions onto the screen itself would reduce the time and disorientation of looking away from the monitor.

• A quasi-intelligent target/fighter assignment might speed up the process of attacking threatening adversaries.

These four features were coded into a training simulator at Brooks Air Force Base.

Once the system was up and running, the research team evaluated the revised interface. Seventeen AWACS weapons directors were brought in for a two-day evaluation period. All the crew members were certified and had an average of one thousand hours following their certification (which required a minimum of five hundred hours with

AWACS activities). Thus, the interface evaluators had an average of one thousand five hundred hours of experience with the existing AWACS interface. The question was, would these highly experienced WDs perform better when using an interface redesigned around a few key cognitive issues?

On day 1 of the evaluation, the AWACS crew members spent two and a half hours getting familiar with the Brooks Air Force Base simulator. After that they practiced using the redesigned interface for four and a half hours. On day 2 they worked with a series of challenging scenarios. For some of the scenarios, they used the standard AWACS interface—the one they were all used to working with during actual AWACS missions. For other scenarios they worked with the redesigned interface. The evaluation study was designed so that performance using the standard, familiar interface could be compared with performance using the redesigned interface for each WD.

The redesigned interface had a strong effect on performance, much stronger than expected given the difference in amount of experience with the two interfaces. The AWACS simulator at Brooks collected data on many different aspects of performance. For example, it measured the time it took for operators to spot a threatening aircraft, how quickly they responded to threats, how effectively they responded to threats, how well they kept track of and protected friendly aircraft, and their overall situation awareness. In all, researchers used seventeen different measures generated by the AWACS simulator. The WDs improved on 73 percent of the measures that reflected how well they were fighting the air battle and on 80 percent of the measures associated with how well they were maintaining good situation awareness. The redesign of the interface had a marked, positive impact on WD performance.

The research team also arranged for a WD subject-matter expert (SME) to examine overall performance using the two interfaces. They compiled all the data onto sheets showing the performance measures for each WD for the standard interface and for the redesigned one. Thus, for seventeen WDs, there were thirty-four data sheets. The researchers gave the data sheets to the expert and asked him to rate the overall performance indicated on each data sheet. To do this, the expert examined all the measures in a given data sheet and then rated the overall performance on a five-point scale where 1 = high performance and 5 = low performance. The WDs' performance using the standard interface received an average rating of 3.8. The performance of the same WDs, using the redesigned interface, had an average rating of 2.82. The overall performance with the enhanced interface was significantly better[1] than with the conventional interface even though the participants were much more familiar and comfortable with the conventional interface.

The AWACS project illustrates how CTA can be used to identify cognitive requirements and then leverage them for system design and for substantive evaluation of usability and usefulness. The project also documents the effectiveness of the resulting interface design. Because of the success of CTA efforts such as this, we have tried to capture the main elements of the process in the method we call Decision-Centered Design (DCD).

The Rationale for Decision-Centered Design

The DCD method was created to use CTA to develop new technologies, including complex human-machine systems (Hutton, Miller, and Thordsen 2003; Klein 1993). The DCD approach involves using CTA methods to specify the primary cognitive requirements and inform the design process. Design teams can use the CTA findings to arrive at better system concepts and features. For that to happen, systems engineers and technology designers need the information that CTA provides, delivered in ways that make sense to them and that fit their schedules, goals, priorities and constraints. One of the challenges for CTA practitioners is to identify ways to influence the design process and to become part of the design team.

But why do we call it *Decision*-Centered Design? As we had found in the development of the CDM (see chapter 5), focusing on people's decision making gives us the most leverage. The DCD designation was adopted in order to emphasize the importance of the key decisions that operators have to make. Once the CTA study has answered questions including, "What are the tough decisions?" "What makes them tough?" and "How do skilled personnel make these decisions?" the researchers can address the basic design issue, "How can technologies help people do a better job with these difficult decisions and cope better with all of the cognitive demands of their work?"

The premise of DCD is that CTA can enable the design team to support the key decisions and overcome the associated sources of difficulty. Asking about critical decisions results in a better understanding of all the important aspects of cognitive work—information requirements, skills, and so on. An open exploration of naturalistic decision making usually encompasses most if not all of the macrocognitive functions. While acknowledging the leveraging potential of the focus on decision making, we also acknowledge that many macrocognitive functions are involved in cognitive work. But the CTA effort does not have to consist of separate investigations into each individual cognitive function or process. We use the DCD designation as an umbrella for all of the aspects of human cognition that are involved in work. In making critical decisions, all of the relevant cognitive functions come into play. People don't make decisions without perceiving and making sense of situations, without noticing problems, without coordinating with others, without setting and adjusting their goals, and so on.

Using Tough Cases as a Basis for Design

Decision-Centered Design is oriented around the tough cases—the difficult and challenging decisions that the system has to support. The focus on tough cases serves a number of goals which we now describe.

The focus on the tough cases and the difficult decisions is intended to ensure that the technology will be designed to be robust and rugged. Many design approaches result in systems that are designed to handle the routine cases, not the tough ones. They are designed around task listings and data flow analyses. As a result, when the work is going smoothly and the standard procedures are being followed, the decision aids and support technology may make the routine jobs easier. Unfortunately, the resulting systems are often brittle and cannot bear being pushed. They are examples of clumsy automation (Koopman and Hoffman 2003) that can actually get in the way during nonstandard events—to the point where operators may just turn off the system rather than trying to wrestle with it. Decision-Centered Design starts with the tough cases; if the system can support these, it generally will take care of the ordinary routines. And if the system is incomplete, the developers typically find it easier to add more functionality for the routine cases than for the challenging ones.

The focus on the tough cases and the difficult decisions is intended to ensure that the development process is efficient. Some researchers argue that unless you study every relevant variable you are not doing a satisfactory job of CSE. Our experience is that sponsors rarely, if ever, have the time or funding to cover the entire cognitive waterfront. To have an impact, researchers have to make priority decisions about how to understand the cognitive work in a domain. We have found that the most direct and efficient starting point is the set of challenging decisions that have to be made in a domain. The key decisions help us map the thinking of the individual operators and the team. Focusing on challenging decisions also allows researchers to provide guidance to design teams about where the highest-impact design elements are.

The focus on the tough cases and the difficult decisions is intended to address the envisioned world problem (Dekker and Woods 1999). Critical decisions serve as a stable point as other trappings of work are changing. Key decisions persist even as software, equipment, procedures, even job roles are changed. Task analyses that are based solely on the study of existing and legacy equipment and/or procedures may result in technologies that only make ineffective procedures and poor interfaces less bad.

The focus on the tough cases and the difficult decisions can allow the researcher to serve as an advocate for the users. One of the shortcomings we have observed in system development projects is that the real needs (the cognitive requirements) of the users aren't described very well and are rarely a focus in requirement specifications procedures. Sometimes they are completely ignored. Other times, there is a brief nod at CTA but no depth of data collection or analysis. Inevitably, this leads to frustration downstream when the system is unveiled or fielded and users realize that it is inadequate, that the new technology puts new burdens on the end-user, and that a major rebuild (often under the guise of an "iteration") is needed. Even when users are queried up front they can sometimes have trouble articulating their strategies or the basis of their expertise. Because their expertise is not understood or reflected, systems are designed in ways that can actually interfere with the strategies they use to handle challenging cases. Critical cues, patterns, and relationships are often obscured. It is all too easy to fall into the trap of designer-centered design, thinking that smart, well-intentioned technologists can put themselves in the shoes of the users and design for them. The resulting systems force the users to adapt to the designer's theory of what users should do.

The focus on the tough cases and the difficult decisions can allow the researcher to serve the clients' needs best. In projects aimed at developing new technologies, we often find that the CTA researchers need to help educate system developers and project managers. Typically, developers are not prepared to understand the cognition of the system's intended users—the practitioners—or even appreciate why such understanding is critical. They usually don't have the background, training, and professional experiences needed to conduct in-depth interviews or grapple with such notions as "cognitive work" or "macrocognition." They may not ask for user inputs because they don't know what to ask. They may ask the wrong questions. They may ask users what they want, which can be counterproductive. When faced with such questions, many users simply identify the snazziest technology they have seen recently. The technological fixes that users recommend are no more likely to result in success than systems based solely on the designer's view. By focusing on tough cases, we can tell the right stories to system developers and program managers to lead them to appreciate why building cognitive requirements into the design might matter.

Another way to describe DCD is to explain what it is not. Decision-Centered Design is not technology-centered. It is not looking to apply the latest and greatest advancement in technology. Too often systems are driven by the need to showcase a new capability, regardless of whether it is needed or useful. Too often, these systems provide excitement and entertainment but are not used because they aren't helping operators make decisions. The goal of DCD is to make effective use of technology and to create useful technologies, not to use the sponsor's funding to advance a pet technology.

Decision-Centered Design is not data-centered. It is not trying to cram anything and everything relevant through ever-expanding information pipelines. Decision researchers have demonstrated that decision quality increases as people receive more information, but only up to a point. That point can be reached pretty quickly (Oskamp 1965). After that point, the additional information has basically no effect on decision quality. But the additional information does increase confidence. The result can be that confidence keeps getting higher ("look at all the data I am using") while quality stays the same, resulting in an overconfidence effect that can be risky.

In the sections that follow we examine the DCD process in detail. We start with an overview and then present a DCD project from beginning to end to illustrate the stages of the process and show how they inform the eventual system design. Advocates of approaches having other focus points (see Hoffman, Feltovich et al. 2002) actually do things very much like what we describe here.

Overview of the Decision-Centered Design Method

The DCD approach consists of five stages:

1. **Preparation** Understand the domain, the nature of the work, and the range of tasks and functions; identify where to focus CTA resources; select CTA methods
2. **Knowledge elicitation** Use CTA methods to conduct an in-depth examination of the key decisions and cognitively complex tasks
3. **Analysis and representation** Decompose data and structure it to identify decision requirements and design leverage points
4. **Application design** Iteratively develop design concepts and application prototypes that support users' decision making
5. **Evaluation** Identify critical measures of performance; evaluate and improve the prototype

Figure 11.1 shows the five stages[2] in the DCD method, along with some steps in each stage.

Inherent in the DCD process is the notion of iterative design. Design and evaluation of concepts, mock-ups, and prototypes, both formal and informal (stages 4 and 5), will suggest refinements, reveal problems that must be fixed, and point toward the neces-

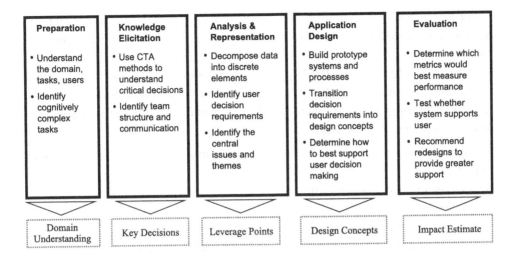

Figure 11.1
Decision-Centered Design (DCD).

sity for additional CTA data collection and analysis. Let's examine what the process looks like in an actual DCD project.

An Example of Decision-Centered Design

In this section we provide a look at an application of DCD from beginning to end. The project was aimed at helping the Navy use information technology to respond rapidly and effectively when a ship has been critically damaged (that is, a casualty event). Fast, effective damage control on Navy ships is critically important—lives depend on it. In this project, data from the initial CTA study were translated into decision requirements and usable implications for design. We used these to generate recommendations for a first-of-a-kind decision support system and HCI.

What problem were we presented with? The Navy directed that ships of the future will have a sharply reduced crew size. Automation can help the crew perform its missions, but what will happen if the ship is hit by a missile or trips a mine and catches on fire? In the past, these circumstances have sometimes required the full complement of crew members. Every single person on the ship has special emergency roles and tasks and has been needed to keep the ship from sinking. If the crew size is reduced by a third of its original number, new types of automated aids will clearly be needed. These aids will have to support the ability to accurately characterize a casualty event, reduce significant delays in containing or controlling fire and smoke spread, restore vital flood systems, and effectively manage the crew members and resources used to control a fire aboard ship.

Miller et al. (2002) used a DCD approach to develop a damage control personnel management system named DC-TRAC (damage control–tracking resources and crew). The DC-TRAC project sought to improve the command and control of handling shipboard fires in the following ways:

- Improving situation awareness of the damage control assistant (DCA) in charge of the fire,
- Speeding up the decision making,
- Enabling better allocation of resources,
- Facilitating workarounds and adaptations to unexpected events,
- Helping the damage control team strengthen their skills and develop expertise,
- Reducing workload, and
- Helping the team gauge the effectiveness of their response.

Stage 1 in the DCD process was to prepare the team. They reviewed documents, studied state-of-the-art damage control systems, and attended a Navy tutorial on damage control.

Stage 2 was to conduct the knowledge elicitation procedures. The team performed more than twenty-five CDM interviews in the field setting with current and former DCAs, damage control investigators, and other damage control experts. They also interviewed instructors at the Surface Warfare Officers School and personnel at the At-Sea Training Group in Norfolk, Virginia. In addition, the team observed damage control incidents aboard the ex-USS *Shadwell*. The *Shadwell* is a decommissioned ship, moored off the coast of Mobile, Alabama, that is specifically configured to experiment with damage control tactics and technologies. The *Shadwell* is the world's largest ship fire research complex. The Navy uses the *Shadwell* as a laboratory. They set controlled fires and study the effectiveness of different technologies and tactics. The DC-TRAC team documented a total of eighteen incidents from these observations and interviews.

In stage 3, the team performed extensive analyses on the data they had gathered. Some of the analyses were detailed, incident-by-incident examinations. Others compiled findings and identified themes across the full data set. Table 11.2 presents a summary look at the major types of decisions gleaned from the observations and CDM interviews.

In a separate set of analyses, the project team extracted the major high-level goals of the DCA, shown in figure 11.2.

Table 11.2
Categorized decisions

Category	Decisions
Ship-wide decisions	Determine whether or not to go to General Quarters
Personnel decisions	Determine how to utilize personnel Determine if we are winning the fight against the casualty Determine which people have the best skills for the job Assess how to coordinate/organize the fight Determine workarounds due to injury Orchestrate safe avenues of approach Determine how/if to monitor personnel actions
Casualty characterization decisions	Determine the scope of the problem Determine what is possible in an incident Characterize the casualty type and severity Determine the cause/source of the damage and risk for progression
Casualty response decisions	Determine where to set boundaries Judge if boundaries need to be modified Prioritize response based on mission and incident severity Determine if primary and secondary boundaries are holding Determine the potential for cascading casualties Determine how successfully the damage is being controlled Determine whether to release extinguishing gases Determine extent of damage and clean up effort
Data integrity/communication decisions	Determine if information is accurate or outdated Determine if information is trustworthy Determine how to keep the CO/bridge informed

Analysis of the data indicated that in many damage control incidents the DCAs were frequently confused about the location and extent of the fire and the location and status of crew members. The strength and importance of these findings suggested that they were good candidates for decision support. Accordingly, the researchers designed DC-TRAC to take advantage of technological capabilities to use sensors and telemetry data to increase the DCA's understanding of the situation.

The results also highlighted several additional aspects of DCA expertise. The skilled DCA:

• Learns about his or her ship by walking around and noting specifics about every compartment—mentally simulating how a problem could spread and identifying potential dangers that might produce a cascade of casualty events,

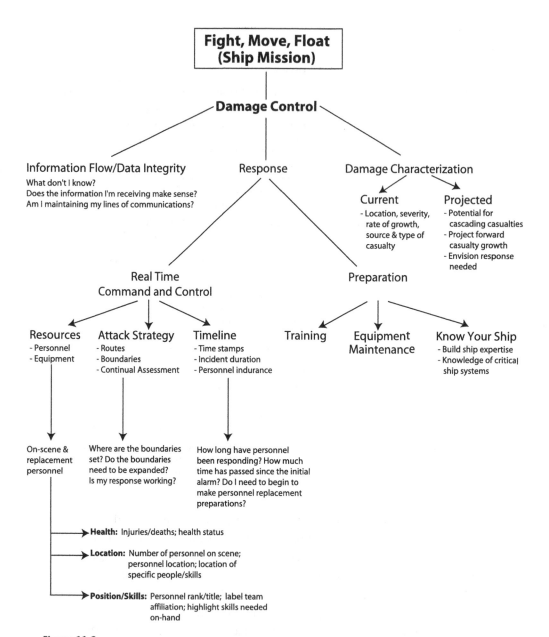

Figure 11.2
Damage control goals.

- Knows how to keep the big picture and not get lost in the details,
- Exhibits proactive behavior for obtaining the information needed,
- Communicates effectively and quickly, and
- Remains comfortable with uncertainty that results from data ambiguity.

The research team created a large set of DRTs to represent the data about the DCA's expertise and situation awareness. The DRTs contained data organized and catalogued by types of decisions, along with cues and knowledge, why the decision was challenging, expert strategies, aspects of information flow, and potential errors.

In stage 4 of the DCD process, the project team transformed decision requirements into design concepts. They identified two primary design elements.

Concept 1: Support the DCA's ability to build and maintain situational understanding of unfolding events by:

- Providing information that makes the scope of the incident more visible.
- Highlighting resource location and status in order to support uncertainty management by helping the DCA answer the question, "What should I be supporting and how should I support it?"
- Bringing the DCA into the loop and up to speed faster by providing key information for establishing situation awareness.

The primary design concept behind DC-TRAC was to allow the DCA to rapidly build and maintain an understanding of unfolding events. The research team achieved this by providing the DCA with a graphical interface that represents the ship (see figure 11.3). The interface allows the DCA to easily see critical situations such as the vertical spread of a fire, the health status of an individual, or the compartment location of a Rapid Response team. The project team's goal was to design DC-TRAC so that the DCA could quickly regain awareness of events by interacting with the system, rather than having to use voice communications that could distract other members of the damage response team.

Concept 2: Support acquisition of DCA domain expertise by:

- Creating a tool that supports the development of skill at making rapid decisions,
- Providing ship knowledge for nonexperts to identify potential cascading casualties.

The project team designed DC-TRAC to support the "know your ship" skill-building function, to raise awareness of the ship's crew (their location, status, and skill information) and to provide a platform for visualizing ship features, crew location, and characterized damage in a single framework. Figure 11.3 shows a screen face for the DC-TRAC system and illustrates the different functions.

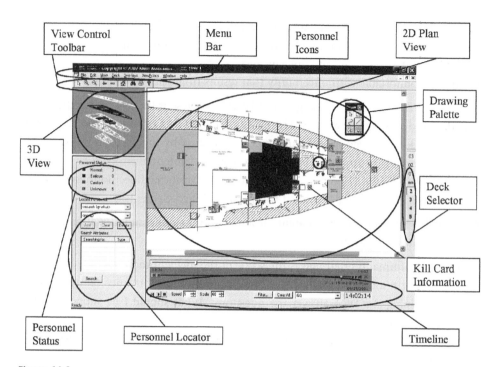

Figure 11.3
The DC-TRAC interface.

As shown in figure 11.3, DC-TRAC is organized around a graphic that provides a spatial representation of the ship's layout. This view lets the DCA easily shift perspective from deck to deck and track the locations and status of the crew members (who carry personnel tags providing telemetry data). The interface includes a damage "palette" to mark the type and location of the damage to the ship (based on information provided by embedded sensors) and various viewing control options. In addition, DC-TRAC lets the DCA use the personnel locator system to search for specific skills or individuals and a health summary window to implement in-depth health status information for individuals. A separate function allows the DCA to review the fire's status and gauge the overall trend of the fire—is it spreading or being controlled? Instructional features of DC-TRAC include a replay capability and an incident timeline to automatically mark time-stamped critical events.

Stage 5 of the DCD process was to evaluate DC-TRAC. The prototype was demonstrated for four days on board the ex-USS *Shadwell* to see how useful it would be for combating an actual fire. Evaluation in live-test situations is different from using staged simulations; the abstract concepts of a design quickly become concrete when they are confronted with the realities and stresses of flames, smoke, and heat.

Test scenarios were prepared by the Naval Research Laboratory to challenge the DCA, the casualty coordinator, and their crew. The DCD team's evaluations included examining DC-TRAC for its impact on decision making, situation awareness, building expertise, managing attention, and detecting problems. Evaluation dimensions for each of the specific cognitive elements were identified and assessed, based on the original CTA findings. For example, the researchers evaluated the DCA's ability to make decisions about covering areas in danger, to prioritize casualties, and to use information effectively and disseminate it to others.

The evaluation showed that the DC-TRAC system was effective for supporting firefighting operations and for training. One of the key benefits of the DC-TRAC system was to help the DCA visualize the damage on multiple dimensions, viewing the different affected decks in parallel, spotting potential problem areas, and understanding the engagement geometry and the vertical spread of the fires. The DCA primarily communicated with the crew via radio, but when communications were down or when the DCA hadn't received radio reports in awhile, the DCA turned to DC-TRAC for location information: "Okay, I've got men on the second deck and main deck...that's where they should be." The evaluation study concluded that the main use of the system was for detecting problems: "Where are my people? Where are they headed? Where's the fire? Where is the fire headed?"

The evaluation also helped the project team see where they needed to strengthen the design. The test and evaluation stage produced a clear set of recommendations and priorities for additional development. Several of the lessons learned in developing the design concepts and features for DC-TRAC have transitioned into damage control analysis and interface usability testing within the Navy's DD(X) program for future ships.

Next we describe each DCD stage more fully and follow the development of DC-TRAC as it emerged from the DCD process.

The DCD Process in Detail

Stage 1: Preparation—Understanding the Mission Context

The first stage of DCD is to gain a big-picture view of the domain. As we saw with the DC-TRAC project, the DCD team starts by coming up to speed as quickly as possible through background preparation. The team might review available materials and gather task and mission analyses. The team members might attend training sessions in the domain or conduct initial interviews and observations as in any CTA. These activities let the CTA researcher perform the framing activities needed to prepare for knowledge elicitation.

Next, the DCD team conducts an analysis of tasks and functions. Before doing any CTA procedure, such as knowledge elicitation, the nature of the work must be sorted out. The function/task analysis might draw on tools and techniques such as Hierarchical Task Analysis (Annett 1996; Shepherd 2000), data flow diagrams (Balzer and Goldman 1986; Woodman 1988), or task diagrams (Klein and Militello 2004). The goal of this stage is to generate an overall perspective on the domain in which the relevant tasks and work are being performed. In the DC-TRAC example, the material in figure 11.2 summarized the information gathered during this initial stage and provided an overview of the damage control function.

The next step is to review the tasks and functions to identify which ones seem worth studying with a CTA procedure. For example, if they had a task diagram the researchers would interview domain practitioners to gauge how cognitively complex the function or task is. They would identify the key decisions that needed to be studied further (Hutton et al. 1997). The researchers conduct this review to flag those tasks or functions where the greatest cognitive challenges exist and where applying CTA will have the greatest impact.

A reason for this step is efficiency. It is impractical to conduct a CTA on every task and function. Such research would take too long, cost too much, and create data overload. If additional CTA data are needed later, they can be obtained, but at this point it is essential to determine how to make the best use of limited resources. Many tasks are primarily procedural; they are performed by following steps and may not require skilled judgment. We see no reason to apply CTA resources to tasks such as these. In the DC-TRAC example, the DCD team informally reviewed the damage control functions and identified the high payoff decisions to study. Table 11.2 lists the shipwide decisions, personnel decisions, casualty characterization decisions, casualty response decisions, and data integrity/communication decisions.

Stage 2: Knowledge Elicitation
The second stage of DCD is to collect CTA data on the functions, tasks, and decisions that were flagged as high priority. This stage is the discovery engine for DCD.

In chapter 2, we identified a great number of methods and tools for collecting CTA data. Depending on the nature of the tasks and the priorities, different blends of methods may be synthesized. The point of knowledge elicitation for DCD is to collect, analyze, and represent the information about key decisions and cognitively complex tasks in order to inform the design and support the design team.

In the DC-TRAC project, an important finding from the knowledge elicitation stage was about the progression of understanding. The researchers found that during an incident the DCA's understanding moved through different "modes." The research team was not aware of any description in the situation awareness literature differentiating different modes, but their CTA research showed the importance of supporting the DCA for prepriming, getting up to speed, building and maintaining situational understanding, continuing to monitor the situation, and handling the maintenance requirements. Which mode the DCA was in depended on the severity of the incident, the success in controlling it, and whether the DCA was primarily pulling or pushing information. In many incidents the DCA would progress through all five modes in order, but often the DCA skipped or repeated modes. This was a key discovery about successive modes of situational understanding that guided the subsequent design of the system. Thus, we see how the discovery process is part of a CTA study.

Because most tasks are performed in teams, we also want to reflect the constraints and costs of teamwork. In order to understand team aspects of decision making, it may be important to perform a team CTA and then document the results with a DRT. Klein (2000) and Klinger and Hahn (2003) have described several team CTA methods. These include:

- The Wagon Wheel method for clarifying roles and functions;
- Team decision requirements exercise for capturing the difficult decisions that teams handle, the contributions of individuals to these decisions, and the teamwork needed to make the decisions; and
- Team Knowledge Audit for clarifying how the team operates, where the expertise is, the intent for forming the team, and the roles and functions of the team members.

Stage 3: Analysis and Representation

The output of this stage is a DRT or other type of data representation that identifies and communicates the leverage points and key decisions. The analysis and representation activities are aimed at capturing all of the primary decision requirements because these are going to inform the design. However, the decision requirements have to be linked to the larger context and the rest of the mission in order to keep track of different goals generated at different levels and to support any trading-off that is needed to cope with goal conflicts. The system design that emerges will reflect this wider field of goals. The design will depend on the way competing goals are traded off.

The representation should enable people, including the researchers, to appreciate how the decision makers are experiencing the task. It should identify critical cues. But merely listing these cues is not enough. The representation has to highlight the relationships between cues and define the critical distinctions between similar cues, because whatever system is designed will have to preserve and magnify the relationships and support discriminations and distinctions, not just the isolated data elements.

If the task involves teamwork, it may be necessary to develop DRTs or other representational formats to highlight coordination issues and other aspects of team decision making.

Stage 4: Application Design

The primary inputs from the CTA research are the various analyses and representations produced in preceding stages. The DCD researchers will also prepare additional inputs, including the individual and team representations that will be more detailed than the DRTs.

The DCD method synthesizes into a set of design concepts what has been learned from the CTA studies, from the system requirements documents, and from review of the current state of the art in information technology.

The DCD process does not reduce the design process into a procedure. Creativity and experience are essential for design (Hoffman, Roesler, and Moon 2004). It is possible, and often important, to create an "audit trail" that links the eventual design back to DRTs and decision requirements. But the audit trail shouldn't overwhelm or drive the design process itself. Attempts to proceduralize the design process often seem to lead to sterile, unimaginative results, such as "There's a requirement here, so the design recommendation is to insert a decision aid using intelligent agents to provide the necessary information." Recommendations such as this can be made almost without thinking, but to what end? If the goal is to generate a thoughtful set of design concepts, then what is needed is thoughtfulness, not procedures.

Optimally, stage 4 is performed as a partnership between the CTA researchers, domain practitioners, designers, and engineers. Despite the drama of springing a radically new design concept on the rest of the development team, and waiting for the "oohs" and "aahs," we find that dramatic unveilings often lead to "huhs?" and "yucks!" The DCD approach depends on partnership and ongoing collaborative effort.

We have found that DCD has to be positioned to contribute to the overall system development process by whatever methods and design processes have been adopted by the organization. Efforts to make cognitive requirements drive the design process lead to frustration and create barriers. The DCD process is aimed at providing a route for decision requirements to enter into system development, not to usurp or replace standard system development approaches.

Stage 5: Evaluation

Any process for building usable and useful information technologies must rely on measures that tap into decision requirements (chapter 14 provides a more detailed account of using CTA studies to design cognitive evaluations). The DCD method uses decision requirements as test and evaluation criteria. The decision requirements identified through CTA can be and should be used to define test and evaluation criteria even before stage 4. The CTA process provides the basis for defining measures up front so they can guide the design process. It is a good tactic to establish decision requirements well before stage 4 so that the cognitive demands are reflected in the design from the beginning. If the metrics are identified after stage 4, they will reflect the system concepts rather than the cognitive demands of the work.

The decision requirements can be used to formalize acceptance criteria and to keep the developer's attention focused on satisfying cognitive demands. The representations created in stage 3, such as DRTs, will influence the evaluation criteria by identifying the decision requirements and what makes them difficult. It then becomes possible to judge technology or human-machine interaction by whether it addresses these difficulties. In short, we can define evaluation measures and metrics that show whether the new system is doing its job of supporting the key decisions on both an individual and a team level.

Once a prototype is developed, measurements can show how well the prototype works and flag weaknesses that can be reduced or eliminated. In the AWACS example we cited at the beginning of this chapter, the researchers performed pilot tests of their prototype interface. They found that each of the design concepts needed to be modified. The lessons they learned during the pilot tests contributed to the effectiveness of the interfaces that were tested more formally.

These five stages constitute the DCD method. The method provides a basis for cognitive engineering so that developers have a process for taking cognitive demands into account.

Costs and Benefits of Cognitive Requirements in System Development

The costs of applying CTA to system design, whether using DCD or other CSE methods, are reasonably clear. Compared to other designer-centered approaches, time and funding are needed early in the design process in order to conduct the individual and team CTA studies and analyze the results. Typically, the research team is short on both commodities. Typically, they are faced with tight schedules and lean budgets. Typically, they are faced with hard and unavoidable pressures—practical constraints that will not go away. It can seem as though incorporating CTA into the mix just makes the process harder, reducing the resources, delaying the schedules, and adding to workload. Why should people—researchers, systems developers, or sponsors—sign on to CTA?

Performance Breakthroughs

Using CTA can improve the quality of the technology that is developed. Klinger and Gomes (1993) demonstrated 15–20 percent performance improvements for the AWACS WD project. This level of performance improvement based solely on more powerful computers or more sophisticated software aids is dubious, at best, given the common experience people have with user-hostile technologies created on the basis of designer-centered approaches. Therefore, a major benefit from using CTA is to achieve performance levels that may be difficult to attain otherwise. However, many technology development projects do not provide any incentive for achieving higher performance levels beyond satisfying the basic criteria for acceptance. For that reason, the program managers focus on satisfying the requirements as quickly and economically as possible. They have no motivation to produce performance breakthroughs.

Avoiding System Breakdowns

Design methods using CTA can help the developers avoid design breakdowns. These are the opposite of breakthroughs—they are the design failures that render systems ineffective or even dangerous. The infusion pump we described back in chapter 10 is an example of when the technology increased the chances for patients to make dangerous self-medication errors. Advocates of CSE have written up many case studies of design failures (Reason 1987; Vicente 1999; Woods 1994), along with principles for reducing the risks.

Klein (2004) has described some of the ways that sophisticated technology can interfere with performance. Developers may be insensitive to the expertise of the decision makers and consequently see the decision makers as people whose job is to operate "their" technologies. This insensitivity would make it difficult for decision makers to gain access to the data they need because the information is fused and smoothed before it ever gets to the user. In addition, the systems can make it difficult for users to form mental models because the users have to conform to the mental models of the people who designed the technology. Systems can make it difficult for the users to spot problems because users are expected to adopt a passive stance and go along with "recommendations" from the computers. Systems can make it hard for users to adapt because the technology is too brittle to tolerate variations. Systems can make it difficult for users to trace the origins of data and analyses because the computers keep their own operations and "intentions" largely hidden from view. Basically, Klein argues that the developers of technology sometimes do not understand or appreciate the way that users think. That is one reason why CTA methodology has developed over the past two decades. But CTA can only work if program managers and technologists are motivated to take cognitive support more seriously.

Time and Money

Using CTA during system design should actually save time and money. CTA should increase the likelihood that designers will be more successful more often in passing test and evaluation reviews, thereby reducing the number of design iterations. Each design iteration can be costly and time-consuming. By using CTA effectively, we believe developers stand a much better chance of satisfying the user panels and gaining acceptance. We make this assertion cautiously, because CSE researchers have rarely had the opportunity to collect the hard data to support this claim. Nevertheless, the growing number of case studies of failures of the traditional procurement paradigm is difficult to ignore. In our own DCD efforts, software developers have been able to produce successful systems in 30–50 percent less time than they expected. Savings such as these would more than offset the resources needed to conduct and apply CTA.

There were several reasons why DCD reduced system development time. First, the software developers did not go down blind alleys. From the beginning, they had a better idea of what was needed, beyond the specifications themselves.

Second, the CTA researchers made sure that the software developers had access to the CTA data. The developers understood how the users were making decisions and performing cognitive work. In some cases, including the DC-TRAC project, the software specialists even went on CTA data-collection trips. These experiences helped the software developers appreciate the role of cognitive requirements and the environment in which the system would be operated. The expense of bringing software developers on these trips was more than offset by the awareness created and the subsequent motivation to create designs that would work in context.

Third, the software developers, armed with the CTA data, were able to press back and suggest alternative approaches to achieve the goals in less time. They had a sufficient understanding of priorities and rationale to become true collaborators.

Fourth, we have established links between DCD and the software development process. The DRTs provide software developers with clear expectations about the level of detail for the enriched task descriptions they need. By viewing the DCD steps, as shown in figure 11.1, they can plan for the amount of time needed to insert CTA studies at various points in the process. Furthermore, they can have a basis for creating enriched scenarios and use cases that will improve the quality of the designs.

Cognitive Test and Evaluation Criteria

We believe that the most effective arrangement to foster collaboration between program managers, technology developers, and CTA researchers is to establish the key decision requirements as test and evaluation criteria. The strategy here is for the sponsors or program managers to spell out in advance the acceptance criteria for the systems and specify the desired changes in cognitive performance. When the program office that sponsors a technology also specifies the desired cognitive requirements up front, then everyone will share the goal of determining how to design a system that supports these requirements. Cognitive measures can be defined in advance of system design and development and can be incorporated throughout the development process.

Improving Design Quality

We believe that any costs of designing for decision making are outweighed by the advantages. Cognitive Task Analysis findings enable developers to design technologies and larger sociotechnical systems that increase performance, reduce breakdowns and brittleness, and save time and funds by cutting down on design iterations. If they do not perform CTA studies, technology developers are forced to guess how their designs are going to improve decision making. Costly systems should not have to depend on guesswork.

Summary

In this chapter, we examined some specific ways to use CTA to generate cognitive requirements for system development and design. We presented a particular approach, DCD, in order to illustrate and describe some general principals of cognitive engineering. Information technology offers enormous potential for improving decision making, but only if the developers of technologies take cognitive requirements into account, not just at the test and evaluation stage, but throughout the development and procurement process.

If you want to train people to do a better job, what exactly do you want them to learn? Sometimes trainees need to learn correct procedures, or they have to remember some facts or details. Sometimes they have to practice a skill so that it becomes automatic. However, in many cases people have to learn new mental models of how something works, or they have to learn perceptual skills so they can make important distinctions. Sometimes people have to learn how to do a better job of managing their attention.

When the point of training is these kinds of cognitive functions, you may want to use a Cognitive Task Analysis (CTA). Simply teaching a list of steps and subroutines is not enough if there is more to skilled performance than following the steps. People need to recognize which steps are important, how to notice that the situation is different from what they expected, and how to adapt the steps. Cognitive training can be more challenging than training people to follow procedures or remember facts.

Cognitive skills can be tricky to train because they are hard to see, to demonstrate, and to describe. For a procedural task, instructors can observe performance to gauge whether the trainees have mastered the steps of a complex action. But how can you tell if a trainee has learned the right mental model and understands the causal connections for that action?

In cognitively complex tasks, the training itself can be difficult to design and implement. This is particularly true when the goal is to help people develop advanced skills. In order to move trainees to higher levels of proficiency, the training usually needs to go beyond practicing component skills. That's why CTA is so valuable—it lets you describe the cognitive skills that underlie real competence and mastery. Klein and Hoffman (1993) have described how tacit knowledge and subtle perceptual skills can be essential for skilled performance—skills that may only become visible through a CTA study.

Cognitive Task Analysis can support training in a number of different ways:

1. Cognitive training requirements   Cognitive Task Analysis can help to identify cognitive training requirements—the kinds of mental models experts have learned and that novices need to discover.

2. Scenario design   Cognitive Task Analysis can help to develop materials for scenarios by using the stories gathered during Critical Decision Method (CDM) interviews. Cognitive Task Analysis has also been used to help trainers design games and simulations that address important cognitive requirements.

3. Cognitive feedback   Cognitive Task Analysis can help to provide feedback to strengthen new mental models and assessments about the effectiveness of training. One strategy is for the instructor to use CTA probes to peek into the minds of the trainees to see what they understand.

4. On-the-job training (OJT)   Cognitive Task Analysis can also improve OJT by letting subject-matter experts unpack what is in their own minds and make their mental models and perceptual discriminations more available to novices. Cognitive Task Analysis can help people in the workplace recognize and share their own skills and expertise and also recognize the skills and expertise of their co-workers and colleagues. The benefits here can go beyond OJT to include knowledge management and the requirement to "train the trainer."

Identifying Cognitive Training Requirements

If we are going to support cognitive performance we need to appreciate it. We need to understand what kinds of cognitive functions have to be strengthened, why people struggle, and how to move them to the next level of proficiency. We need to know how people make sense of situations, where they get confused, why they get stuck with flawed mental models, and what kinds of relationships experts see at a glance that novices don't even notice. The CTA stage of identifying training requirements is at the heart of cognitive training.

Here is an example of how to use CTA to describe the types of patterns firefighters needed to learn.

Example 12.1
Collapsing Buildings

The National Emergency Training Center (a component of the National Fire Academy) requested a CTA to improve their training manuals. They wanted us to provide some general decision-making principles, using the recognition-primed decision model (Klein 1998; Klein 1989a). However, we didn't see how general principles would help in such training. We are usually skeptical of attempts to boil complex cognitive skills into steps that can be captured on a page and assigned for memorization. Instead, we agreed to use CTA to examine specific skills that firefighters needed to make better decisions when entering into unsafe conditions.

We reviewed a draft of a new manual that the National Fire Academy was preparing to release. We identified critical judgments and patterns that were worth probing with CTA methods. We modified the manual so that it provided more useful material on these judgments and decisions (Klein and Wolf 1995).

For example, the original training materials tried to provide guidance on the need to exit a building if there was a chance of collapse. However, the guidance did not appear to be particularly helpful or informative:

Look and listen for signs of collapse. Newer, lightweight construction makes predicting building collapse difficult or impossible. Constantly monitor the building condition and pay attention to all signs of collapse.

This advice didn't seem useful. The firefighter is told to be careful and monitor, but not what to look for. Even worse, the task of monitoring may be impossible with new kinds of construction.

After collecting CTA data, we augmented these materials by adding the following information:

Unless you know otherwise, all construction should be considered to be lightweight. This type of construction is the most dangerous to the firefighter from the collapse-potential point of view. Assume lightweight until you are able to confirm some other, more substantial type. In addition, you must also remember that older construction technologies do not ensure safe operations.

In lightweight construction, such as parallel-chord wood truss, plywood I-beams, or pitched truss, collapse has been known to occur in as few as five minutes after the fire has involved the assemblies. For steel-bar joist truss, collapse has occurred in as few as nine minutes.

Collapse indicators may include creaking or cracking sounds, but during a fire you may not hear these. There are other clues to look for:

- Building distortion (twisting, leaning) must be recognized.
- Horizontal cracks in drywall on interior walls may indicate that the floor is sagging and pulling away from the wall assembly.
- Horizontal cracks in exterior brick may indicate wall failure. Remember, many times the walls hold the floors up. Realize that if the walls fail, the floor will also collapse.
- Vertical or diagonal cracks, or bowing of brick walls need to be recognized and monitored. Again, wall failure could follow.

The length of time the fire has been involving structure-bearing members must be monitored. Unprotected metal members ordinarily fail rapidly when exposed to high heat or direct flame contact, even for short periods of time.

In this example, we didn't try to develop a separate set of cognitive or decision-making materials. Instead, we worked with the National Fire Academy to embed the decision training within their existing course structure. Our embedded materials told the firefighters what to look for, and not simply to be careful. That's the point of cognitive training. The CTA helped us to identify the cognitive training requirements and to inject them into the existing scenarios. We also demonstrated the benefits of making explicit the subtle judgments of the experts.

Scenario Design

One of the cornerstones of training is to provide practice and feedback. When trainees face challenging scenarios, they can practice making decisions and judgments and performing other cognitive functions. The training development team must therefore find ways to design scenarios that demand those cognitive skills.

It is not enough to design difficult scenarios in which the workload gets higher and higher—so-called Armageddon scenarios that are typical in many settings. Often, cognitive skills matter most *before* the action ratchets up, to spot the early signs of problems, to make resource allocation decisions that will affect everything that follows, and to make staff assignments that will reduce coordination costs when the pace of activities increases.

A project to train Marines to make better decisions illustrates how CTA can be used to design scenarios (McCloskey et al. 1998). The Marine Corps was preparing to conduct an exercise called "Hunter Warrior." The colonel in charge of the exercise realized that squad leaders, young lance corporals and sergeants, were going to act as forward observers without any direct supervision from their supervisors, the platoon leaders who usually provided them with guidance. These young Marines were going to have to make decisions on their own for the first time since they joined the Corps.

The colonel asked if we would be willing to teach them how to make decisions. As with the National Fire Academy, we declined. We explained that there aren't any generic skills that make people better decision makers. Instead, people acquire more patterns and build more sophisticated mental models. That is how they are able to quickly recognize what to do in situations. However, we did agree to try to help the squad leaders come up to speed more quickly—to gain skills and knowledge that they could use while making decisions.[1]

Example 12.2
Turning Young Marines into Decision Makers

First, we identified the kinds of judgments and decisions the squad leaders were going to make in the Hunter Warrior exercise. We didn't want to teach them general principles of decision making. That kind of decision checklist approach isn't useful out in the field. We had to understand the specific types of judgments the squad leaders were wrestling with.

Next, we clarified why these judgments and decisions were difficult. We created a set of decision requirements tables that guided the way we designed the training.

The third step was to construct scenarios—decision-making exercises that were specifically tailored to the Hunter Warrior mission. We considered using existing decision-making exercises that had already been developed for the Marines (Schmitt 1995, 1996). These exercises would have been interesting for the trainees, but we decided not to use them because they didn't cover the cognitive training requirements for Hunter Warrior. Instead, we designed new scenarios to challenge the squad leaders' judgments, beliefs, and mental models, and to help them acquire experience in rapidly making tough decisions under ambiguous conditions.

Marines are already familiar with decision-making exercises. These paper-and-pencil exercises are an integral part of every officer's training. In addition, the Marine Corps Gazette publishes a tactical decision game every month—a decision-making exercise that portrays a situation map and poses a dilemma. The Marines have learned to run these exercises in groups with a facilitator explaining the dilemma and the trainees trying to find a solution.

To train the squad leaders we administered the exercises in small group settings. John Schmitt, who popularized tactical decision games (TDGs) in the Marines, modeled techniques for running and facilitating the scenarios. The company training officers did the bulk of the training, either out in the field while everyone was waiting around, or back in the barracks.

We also used CTA methods to help the training officers provide feedback to the squad leaders. We prepared a set of CTA-type facilitation probes for the company training officers to use to help the trainees reflect on how they made decisions during the scenarios, which cues they had relied on, and which cues they should have been monitoring more carefully.

During the actual Hunter Warrior exercise the squad leaders did much better than anyone had expected. They found their way to their designated locations and worked as forward observers to make radio calls describing the situation out in the desert to battalion and regimental officers. The squad leaders called in artillery and air strikes where needed and handled a variety of difficult assignments.

Afterward, when they were asked how they did so well, they gave a lot of credit to the decision training program. The training officers in the company also found that the use of CTA probes helped them appreciate what the squad leaders knew so they could identify the squad leaders with good judgment skills.

As a result of this project, the Marines developed a formalized program for noncommissioned officers (Klein Associates 1999) to help them come up to speed more quickly and to make better and faster decisions. We prepared a syllabus for instructors—senior noncommissioned officers—to identify cognitive training requirements, to use cognitive training scenarios, and to conduct cognitive debriefs as a means of helping trainees build decision-making skills.

Cognitive Task Analysis studies have helped us construct useful scenarios in many different settings. For example, Klein (2004) described how to use CTA to improve the decision making of business executives. As with the military applications, the executives were enthusiastic about the scenarios we designed that reflected real cognitive challenges. The executives disliked the generic business games and scenarios they were usually given in seminars. In contrast, the executives appreciated the CTA-based scenarios taken from their own corporate history, which provided additional context and more relevant and more difficult dilemmas.

Crandall and Calderwood (1989) used CDM interviews to develop training materials for nursing students to teach them the subtle cues and patterns of newborns who are in the early stages of sepsis. The nursing students learned the cues and patterns and maintained significant recall two weeks after the training sessions.

Scenarios drive training exercises. They can also drive the design of games that teach decision-making skills. The following example illustrates how CTA was used to shape a computer game and how the researchers modified the decision requirements tables (DRTs) to make them more useful.

Example 12.3
Antiterrorism Computer Game for the Marines

The Marine Corps sponsored a project to teach antiterrorism skills using a computer game: The Anti-Terrorism Tactical Decision Simulation. The program managers realized that the game was only going to be as good as the scenarios. The managers commissioned a CTA study to identify the key decision requirements for antiterrorism decision-making skills and to guide the scenarios.

The project was conducted by CHI Systems and Klein Associates. The researchers used the CDM to conduct CTA interviews with fourteen security supervisors—Marines with antiterrorism skills and experience. The researchers obtained a rich set of incidents, analyzed these interviews and represented the results in DRTs, with rows for each decision or assessment and columns for the critical decisions, the triggers for those decisions, the reasons why the decision was difficult, the factors that had to be taken into account, the strategies used, and the teamwork required.

But the research team went further than just representing their data in columns and rows. They realized that game developers wouldn't readily appreciate the material in the DRTs without knowing the story behind the table. In order to make this material come alive, the researchers supplemented the DRTs with one-page narratives describing each of the incidents they had elicited in the CDM interviews.

And the researchers took another step—they created a DRT for each incident to provide more depth than just one overall table summarizing the incidents. They wanted to make sure they provided the game developers with the right context and not just a list of bullets.

Usually, DRTs are constructed for decisions and not for incidents. One of the reasons why the researchers made a DRT for each incident was because the program managers were interested in the set of "triggers." They wanted to know what triggered the individuals to make a decision so they could build those triggers (cues) into the game. The triggers related to recognition-primed decisions. This was the first time that "Triggers" was used as a category in a DRT, and it was a specific request from the program manager. They wanted to know what triggered each decision; the researchers couldn't generalize each decision/assessment because the triggers were context dependent.

These detailed DRTs showed the role of the decision maker during the incident, a one-sentence summary of the incident, a short description of the challenge, the duration of the incident, the training standards the incident covered, and the nature of the threat. The researchers also made sure that the designers could readily refer back to the one-page incident accounts which provided more context and details. Thus, the values of the representation were in the organization of the DRTs (especially the triggers), the preparation of a DRT for each incident, and the addition of narratives for each DRT in order to provide context to bring the DRT alive.

The researchers additionally prepared a set of recommendations for game design developed from the analyses of the CTA interviews. These recommendations were general issues that emerged in the DRTs, such as ways to increase the cognitive authenticity of the games.

The result was a successful transition. The program managers reported that they used the CTA results and recommendations to develop the computer game. They have also stated that the CTA data and design requirements directly shaped the game that is currently being produced by Destineer Studies. One project manager said that this was the first time they had used CTA data as inputs into a tactical decision simulation—and that this is clearly the way it should be done.

Here is their description of the product, edited to remove all the acronyms (U.S. Marine Corps 2003):

The Anti-Terrorism Force Protection Tactical Decision Simulation: This system under development for the Marine Corps Security Forces Battalion is a PC-based, fast-paced, and tactically realistic computer-based simulation. The system will provide training for armed, antiterrorism and physical security personnel involving the use of deadly force to protect designated installations. The intent is for students, or the training audience, to be presented with a tactical situation for which they develop a plan. The students will then war game their plan using the simulation to provide feedback. Repeated simulation play will enhance their skills. The simulation can be played in a competitive free play mode to develop combat decision-making exercise scenarios where planning is done prior to the simulation, then simulation data is used to provide feedback.

Cognitive Feedback

After a practice session, such as a scenario, one of the most valuable training opportunities is to reflect with the trainees about what just happened and how they understood it. Here is another place where CTA can be helpful: Cognitive Task Analysis methods can be used to find out how the trainee made sense of the scenario and to spot ways that a trainee may be confused.

For example, we used a DRT to structure the debriefing in a decision skills training workshop with Navy helicopter pilots at Naval Air Station, Jacksonville Florida. One of the workshop activities was to present a decision making exercise (DMX) and then have the group generate a DRT for the tough decisions and various courses of action. The DMX presented a situation that forced the pilot playing the exercise to make a decision. The dilemma involved weighing the risks of flying in severe weather against the risks of waiting for the weather to clear and possibly preventing a critically ill patient from getting timely care at a land-based hospital.

Almost all of the pilots chose the same course of action—to take off in the weather conditions described in the DMX. The workshop included officers with a range of skill levels, and the answers were uniform. However, what we discovered as they built the DRT was that the rationale for the courses of action varied greatly. Thus, the class discovered that a young aviator, when describing his mental model for determining whether to take off, failed to account for the pitch and the roll of the ship. The way he described executing his course of action could have very well led him to crash into the side of the ship or in the water on take off. The more senior officers in the room jumped in to help him break down his faulty assumptions and mental models.

In another training program with Marines at Camp Pendleton, we had an opportunity to accompany noncommissioned officers on maneuvers. Afterward, we used a CDM format to conduct the debrief. We asked the Marine leading the unit in the exercise to create a map of his movements, and then to identify the critical judgments and decisions at various points in his progress over the terrain. For each of these decision points we probed the nature of the decision, what he took into account, and, with hindsight, what he should have been noticing earlier.

A project at Clarian Health Partners, Indianapolis, offers yet another illustration of how to use CTA to obtain feedback. The Clarian project was designed to embed knowledge elicitation skills within their organization to support their Safe Passage Program across multiple hospitals. The Safe Passage Program is designed to foster patient safety by designating a nurse-representative for each patient care unit. One of the responsibilities of Safe Passage nurses is to be prepared to identify issues and problems that have the potential to compromise patient safety, including errors and near-misses. We designed CTA workshops to teach knowledge elicitation skills to the Safe Passage nurses so they could do a better job of debriefing members of the healthcare team in the event of a patient safety incident. In the workshops, nurses gained experience in translating statements like "the patient doesn't look good" into richer, more detailed statements that more accurately expressed knowledge about the patient and better communicated the patient's condition and needs to other medical personnel. Nurses learned how to use CTA to explore the "story behind the story." The lessons from this workshop are being integrated into Clarian's Safe Passage training for all nurses.

By using CTA methods in conducting debriefing sessions, instructors can learn more about the mental models of trainees and also give others in the group a chance to learn.

On-the-Job Training (OJT)

In most organizations, most of the learning happens via OJT. The knowledge and skills of experienced personnel is leveraged to bring newer workers up to speed. The potential impact of high-quality OJT is enormous. As much as 80 percent of training provided to adult workers in the U.S. is OJT. In business, industry, health care, and the military, OJT is often the vehicle for how people figure out how to do their jobs.

Yet this form of knowledge transfer is usually ignored and left as an accidental by-product of co-worker interaction. A lot of OJT is completely unstructured—there are no learning goals, training requirements, or formalized assessment procedures. New employees or employees moving into different jobs are directed to observe and shadow an experienced worker for some amount of time—an approach referred to in the training community as, "go sit by Nellie/go follow Ned." When OJT is more structured, workplace-based training is often a combination of observation and shadowing of experienced workers and classroom sessions; structured OJT programs sometimes also involve proficiency testing or a certification process for trainees.

There are many positive aspects of OJT: it offers learning in context, the chance to observe how the job is actually done, and opportunities to interact with and model experienced workers.

What's not so great about a lot of OJT is Nellie and Ned themselves. A lot depends on Nellie and Ned's abilities to articulate how they know what they know and a lot depends on their skills as instructors/coaches. In most work settings Nellie and Ned have had no training in either function. Moreover, our CTA studies have found that the assumption that Nellie and Ned have these key skills and know how and when to use them is a risky one.

Experienced workers, the Nellies and the Neds, usually are eager to share what they know. However, they have trouble articulating the subtle aspects of their expertise. They struggle to explain their intuitions—their perceptual skills, how they make sense of situations, and the mental models that help them perform at a high level.

When we began studying OJT in the mid-1990s, we were surprised at how little information there is about what constitutes high-quality OJT. Wide-ranging reviews of literature on OJT and related topics (e.g., adult learning, situated cognition) provided many ideas about what OJT should be, but surprisingly few accounts of actual OJT programs or descriptions of the skills and knowledge that skilled OJT providers possess (Crandall, Pliske, and Zsambok 1999; Zsambok, Crandall, and Militello 1994).

A series of CTA studies conducted with OJT providers and one-on-one instructors began to fill in the picture of how OJT happens and what is needed for it to be done well (Crandall et al. 1992; Zsambok et al. 1996). We worked in a number of different work settings: critical care nursing, national guard armored units, retail sales, music instruction, and public utility service to understand what happens when one person is tasked with getting another person prepared to function competently in the workplace. These projects revealed how OJT happens, what kinds of skills OJT providers need to have, and how to foster high-impact OJT by helping OJT providers and trainees work collaboratively.

Based on findings across these CTA projects, we developed a description of the key functions that skilled OJT providers perform. They are described in table 12.1.

These initial CTA studies of OJT have provided a solid research foundation and guiding framework for workshops and training programs designed to help organizations do a better job of providing high-impact OJT. We have carried out a number of OJT workshops, including programs in the retail setting, health care, fire departments, and the military.

An example comes from a program developed for the U.S. Navy (Pliske et al. 2000). It was designed to take advantage of the wealth of expertise that exists on board Navy ships and leverage it to enhance shipboard training. The Navy wanted a tool to help shipboard trainers share their expertise more effectively in one-on-one training situations. The research team designed a two-stage OJT program. First, a CTA study was conducted for a given specialty to identify the set of cognitive demands for the position. Second, the CTA findings were represented in a variety of materials developed for workshops for both OJT providers and trainees. The OJT program treated the instructor and the learner as partners who worked together during the training process. They had a mutual goal to make the most of opportunities to transfer domain expertise. Since the focus was on expertise and transferring subtle skills and tacit knowledge, CTA tools were almost essential.

We have provided a number of illustrations and examples of cognitive training and learning. The following case study goes into greater detail about one of the important aspects of applying CTA to training—how to represent the findings of a CTA study in a DRT and how to use the DRT to design scenarios.

Using CTA to Develop Training: A Case Study

Years before Operation Iraqi Freedom, the military realized that future combat was likely to take place in cities, rather than in open ground. Events such as Chechnya, Mogadishu, Berlin in World War II, and Hue City showed that urban combat was likely to incur very high casualty rates. In response to these realizations, the Army and the Marines constructed realistic urban battlegrounds in order to practice urban combat skills, a decade before they needed to apply these skills in Baghdad and Fallujah.

But mock city terrain is not enough; the Army needed to understand the cognitive challenges of urban combat. Therefore, the Army sponsored a project to use CTA to

Table 12.1
Indicators of OJT skill

OJT Function	Skill Components
Sharing expertise	Providing instruction that goes beyond what is available in a training manual or what is involved in basic procedures. May include sharing experienced-based knowledge, skills, heuristics, "job smarts," and judgments. May provide training in how to detect anomalies, recognize opportunities, anticipate and prevent problems, and compensate for errors.
Assessing the learner	Conducting initial and ongoing evaluation of trainee's performance level and diagnosing barriers to expected progress so that instructional method and content can be fitted to current and future training goals.
Setting goals	Setting realistic learning and performance goals. Involves making explicit to the trainee the overall training goals and plan for attaining them and the link between current training activities and overall goals.
Instruction and coaching	Tailoring teaching and coaching practices to the individual's current performance level and training goals. Involves flexibly adjusting or switching training techniques that aren't working.
Promoting ownership	Promoting engagement by offering opportunities for trainee to actively participate in his or her learning and skill attainment. May include mutual goal setting, a collaborative approach to assessment, and individualized instruction.
Setting climate	Creating and maintaining a climate that is conducive to learning—one that is open, supportive, and nonpunitive and promotes frank disclosure from both OJT provider and trainee.
Providing oversight	Guiding the trainee's learning process, and proactively engaging in doing so (rather than automatically following a set of instructional procedures). Requires that the OJT provider develop and maintain a "Big Picture" by standing outside the training process and reflecting on how activities fit together and impact the trainee.

understand the critical elements of decision making in urban combat (Phillips et al. 2001). In this section we describe that project in some detail because it illustrates what a cognitive skills training program can look like.

Our approach focused on helping platoon leaders accomplish the task of clearing a building—taking it over and making sure no adversaries remain. We conducted a CTA of the task and used the CTA findings to develop training recommendations for new and inexperienced platoon leaders. The steps of this project closely parallel the Decision-Centered Design stages described in chapter 11.

Link the key decisions to the mission context The goal was to understand the nature of urban combat and to put it in perspective of military operations. During this step we identified a key aspect of urban combat—clearing a building—as the one to address in our project because it was so difficult and dangerous. Soldiers had to storm a building and counter any threats without knowing whether there were adversaries inside, without knowing the interior layout, without knowing if mines or booby traps had been set, and without knowing if snipers were waiting for them.

Perform CTA on the key decisions We interviewed Army Rangers and other specialists who had experience with actual combat, particularly with building clearing operations in city environments (Phillips et al. 1998). We used the Knowledge Audit to provide an overview of the decision requirements and CDM to compile a set of incident accounts. Results of both methods were used to define decision requirements.

Team CTA We uncovered several important training requirements for team coordination during building clearing, and we added these requirements to the training objectives.

Develop scenarios Because a CDM interview elicits specific incidents, we could readily transform these incident accounts into training scenarios.

Evaluation The final step was to evaluate the training program. We performed this evaluation at the United States Military Academy at West Point, New York (Pliske et al. 2001). The West Point cadets reacted enthusiastically to the training, anticipating that they themselves might have to provide leadership in urban combat in the future.

In the section that follows, we describe in detail the link between performing the knowledge elicitation, constructing the DRTs, and using the DRTs to create scenarios.

We conducted an initial round of interviews with nine Army Rangers whose experience included a wide variety of roles, positions, and experiences. The three non-commissioned officers we interviewed had a total of fifty-seven years of experience implementing orders and carrying out missions in places such as Somalia, Germany, Vietnam, Nicaragua, and Grenada. The captain and lieutenants were able to discuss building-clearing operations from a platoon leader's perspective. The retired colonel supplemented the data collection by providing information and constraints from the perspective of a platoon leader's superior. In a follow-on study by Phillips et al. (2001), we interviewed additional Army Rangers to validate and extend our findings.

We developed a set of Knowledge Audit probes to describe the nature of perceptual, diagnostic, metacognitive, recognitional, and compensation skills for this particular task and skill set. The strength of the Knowledge Audit is that it enabled us to survey the nature of expertise in urban combat environments rapidly. Here are the Knowledge Audit probes we used:

- What is important about the big picture when clearing a building?
- If the platoon leader had to turn over command to a subordinate, what would he tell him?
- What are the major elements a platoon leader has to know and keep track of?
- When clearing a building, have parts of the situation ever "popped out" at you and you noticed things that others didn't catch?
- Can you think of a time when you noticed an opportunity to do something better?

The CDM probes centered on specific, personally experienced incidents in which the interviewees felt especially challenged while clearing a building. The incidents anchored the interviewees so they could speak in specific terms versus describing a generic building-clearing operation. The people we interviewed could recall specific cues, judgments, decisions, challenges, expectancies, and leverage points because the missions we examined were so difficult. Here are some of the CDM probes we used:

- What information did you actively seek to make your decisions?
- What cues did you notice that a less experienced leader would not?
- What did you interpret those cues to mean?
- Describe your situation awareness at different points during the incident.

Using the Knowledge Audit and the CDM findings, we identified eleven key decision requirements. These decision requirements became the cognitive training objectives. Six of the decision requirements were specific to the task of clearing a building. They are determining how to: secure the perimeter of the building, approach the building, enter the building, clear the building of hostile combatants, evacuate the building, and maintain security after finishing the job. Each of these decision requirements has its own challenges and skills. We prepared separate DRTs for each one.

The five additional decision requirements were more general. They were skills that the Army officers needed in all phases of clearing a building but that they also needed for other types of missions. The task-independent decision requirements include:

maintaining the enemy's perspective (thinking like the enemy), leading subordinates, maintaining situation awareness, applying rules of engagement, and anticipating future developments.

Table 12.2 is a portion of one of the DRTs that we developed for the mission of entering a building:

We transformed what we learned from the CTA to create training scenarios. The following illustration centers around the decision requirement of determining how to secure the perimeter of the building. Here is the background for the scenario:

You are an Army platoon leader of XXX UNIT. Your unit has been operating for two weeks as part of a U.N. task force providing security for humanitarian relief efforts. Sonala is a Third World country torn by civil war, the result of which has been a total breakdown of the country's infrastructure creating widespread outbreaks of disease and starvation. The local populace is rapidly growing weary of the terrorists and are no longer hiding them. Enemy infantry forces have infiltrated and have recently begun occupying defensive positions around the capital city of Mondishu in an effort to seize control of the city and assume power. The capital city contains approximately 50 buildings and has a population of about 1000 people. Intelligence reports that the enemy is operating in 2–4 man assault teams on the outskirts of the city, occupying some of the perimeter buildings. They are equipped with AR-15s and grenades. They seem to have little mutual support between buildings, but do have several sniper teams in the area.

Within the decision requirement of securing the perimeter (in table 12.2), there are several critical decisions that must be practiced. The platoon leader must know, for example, how to seal off the area and where to place his security assets. The DRT for securing the perimeter includes these critical decisions and suggests why they are difficult to make, while identifying the cues and factors experienced soldiers use in making such decisions. Using this information, we constructed scenarios that force participants to deal with these decisions. Here is an example:

Your platoon has been assigned to clear a building in the war-torn city. You have planned to secure the perimeter by placing your support unit in a partially walled-in courtyard northwest of the building. However, upon reaching the area around the building you find that civilians are running frantically all around the building, and several currently occupy the courtyard. Your support unit will have to provide security from a different location. To make matters worse, you know that enemy snipers could strike at any time in this inflammatory environment, so you will need to get your units in place quickly.

The cues that the SMEs identified as critical for determining how to place their security assets included:

- Whether foot paths branch off the streets
- Known and suspected enemy locations
- Whether structures can provide concealment or interfere with firing

Table 12.2
DRT example: Determine how to secure the perimeter

Critical Decision	Why Difficult?	Cues	Factors	Strategies/Aspects of Expertise
Determine how to seal off the area	▪ Range of factors to consider ▪ Need to recognize potential problems	▪ Proximity to other buildings ▪ Opportunities for cover and concealment ▪ Enemy activity in area ▪ Civilian activity in area	▪ Enemy capabilities to engage ▪ Intensity level of the conflict ▪ Civilians' feelings toward enemy vs. toward us ▪ Size of the area	▪ The goal is to prevent people from entering or exiting the area ▪ If it's a high-intensity conflict, the unit will probably be trying to clear more than one building; therefore, they will need to secure a larger area
Determine where to place security assets	▪ PL cannot be sure about the makeup of interior walls ▪ PL might give away his attack plan if the support by fire is too close to the building being assaulted	▪ Whether streets are singular ▪ Whether footpaths branch off the streets ▪ Enemy locations ▪ Whether structures can provide concealment or interfere with firing ▪ Perceived best angle to support fire ▪ Anticipated layout of the building ▪ Windows in the building	▪ Areas you want to cover ▪ Effective ranges of weapons ▪ Ability to conceal support element at various locations ▪ Angles of fire ▪ Enemy weapons	▪ Mission success is largely dependent on the support by fire position; it will make or break you ▪ As a general rule, leave 2–3 window lengths between supporting fires and point of entry ▪ Give yourself leeway with regard to angles to make sure you avoid fratricide

- Optimal angles to support fire
- Anticipated layout of the building
- Windows in the building

Using information from the DRT, we can introduce a contingency, an unexpected event that forces the decision maker into action. This event may be guided by the "Why Difficult?" column of the table. For example, suppose this particular task of securing the perimeter is made especially difficult when the rules of engagement state that under no circumstances may civilians be mistreated. This information could be used to create the following quandary within the DMX:

The civilians around the building are now beginning to form into an organized mob. They are refusing to leave the area and are starting to pick up sticks and rocks. Suddenly, one of the civilians throws a rock at your 1st squad leader, the leader of the squad designated to provide security. The squad leader has been knocked unconscious. Seeing this, one of your platoon members strikes the rock-thrower, knocking him to the ground. This angers the crowd even more, and as you wonder how you are going to solve this problem, your company commander calls you, instructing you to immediately get that building cleared. What do you do?

Our CTA research also enabled us to define ways to alter the difficulty of a scenario. These guidelines allow a scenario to be adapted and made more or less challenging. Most of the entries in table 12.3 apply to nonmilitary settings as well as military ones.

This project was a deliberate showcase of how to use CTA studies to guide instructional design, from specifying the training objectives to developing the training scenarios. The research team also developed a CD to prepare instructors who would facilitate the training scenarios (Klein Associates 2001).

Jim Staszewski has provided a forceful demonstration of how CTA can produce effective training. Staszewski's goal has been to improve soldiers' abilities to use handheld devices to detect landmines by teaching them to use the detection strategies of expert demining specialists.

Staszewski (Staszewski 2004; Staszewski and Davison 2000) used a variety of CTA methods, including quasi-naturalistic observation, verbal protocol analysis, and interviews, to identify specific pattern-matching skills and search strategies. Based on the insights provided by CTA, he developed a training program that combined instruction and practice. The program was aimed at developing component skills that would allow soldiers to employ the demining strategies and behaviors of expert demining specialists.

In an initial demonstration of the program, soldiers trained for 12 to 15 hours. Their detection rates were exceptional, ranging from 90 to 97 percent. They achieved these very high rates of success even for the most difficult mines to detect (small,

Table 12.3
Dimensions of DMX complexity

	Basic DMX	Advanced DMX
Level of uncertainty	• Communications are clear • Little or no ambiguity in scenario description or background • Nature of situation is known • Players in the situation are known • Mission is clear • Higher intent is clear • Superiors are readily available	• Mission statement seems to lose appropriateness • Higher intent is vague or missing • Superiors are unavailable • Nature of situation is unknown • Players in the situation are unknown • Capabilities of other players are unknown • Communications are lost
Subtlety of cues	• Cues are clearly presented • Cues form a clear, easily recognizable picture • No irrelevant cues are presented • Few cues needed to form accurate representation of situation • Little inferencing needed to interpret cues	• High number of cues presented • Cues lead to multiple interpretations of situation • Cues are fuzzy • Cues not experienced first-hand • Multiple cues needed to form representation of situation • Significant inferencing needed to interpret cues
Organizations involved	• Within team or unit	• Interaction required with other units, other services, other agencies and organizations
Complicating events	• No unexpected events • Equipment functions properly (comms, vehicles, etc.) • No casualties • Team performs as trained • Weather doesn't preclude normal operations • Simple terrain • Little time pressure (in situation, but in practice, time limit still enforced)	• Unexpected events • Equipment malfunctions or is damaged • Casualties • Inexperienced team members • Weather adversely affects operations • Complicated terrain • High time pressure (situation is developing rapidly)

Table 12.3
(continued)

	Basic DMX	Advanced DMX
Resources available	• Unlimited supplies • Transportation is straightforward • Unit is at full strength and well-trained • Morale/readiness is high • Reliable communications gear	• Limited supplies, given the current situation • Implied need to conserve supplies, given uncertainty of future events • Transport capabilities not ideal for situation • Unit at or near lowest acceptable operational strength • Morale is low; fatigue is high • Comms gear only works intermittently
Situational demands	• Non-emergency situation; action not immediately required	• Emergency situation; immediate action required
Operational constraints	• Rules of Engagement (ROE) are unrestrictive • ROE is simple and straightforward • Freedom to initiate any action without coordination/permission from other agency	• ROE is restrictive • ROE is ambiguous • ROE becomes obsolete in course of mission • Requirement to clear actions prior to implementation
Complexity of mission	• Single task	• Multiple simultaneous and/or sequential, linked tasks

antipersonnel land mines, with low metal content). Prior to training, detection rates for the most difficult mines had ranged from 10 to 20 percent, so the improvement in performance was dramatic. In another study, six platoons of combat engineers (approximately 180 soldiers) who were scheduled for immediate deployment abroad received the training. The only change in this study was to reduce the training time to about one hour per trainee. Pre-training performance was only 20 percent accuracy detection. Post-training performance reached approximately 80 percent detection rates.

In a follow-up project, Staszewski applied his expertise-based training approach to soldiers' use of a new detection technology, the Handheld Standoff Mine Detection System (HSTAMIDS). After nine years and a $38 million investment, the U.S. Army was considering canceling the HSTAMIDS program, primarily due to substandard performance of the prototype in operational tests. U.S. Army Red team attributed that substandard performance largely to training deficiencies. But the training approach based on CTA showed dramatic impact on soldiers' abilities to use the new technology to detect mines, including very low metal explosives that are the most difficult to detect. The U.S. Army has adopted the training approach and now uses it to prepare soldiers for countermine operations.

These studies clearly demonstrate the value of using CTA to reveal expert knowledge and skill in order to use it to create an effective training program. The critical importance of landmine detection in many areas of the world serves to underscore the potential and broad value of CTA.

Summary

CTA is central to cognitive training. If you want to improve cognitive functions such as decision making, sensemaking, problem detection, replanning, and so forth, you will need to understand how these functions are accomplished and where people struggle and run into difficulties. In this way, you can define cognitive training requirements.

The primary applications of CTA for cognitive training are to identify and define the training requirements, discover the basis of skills, the source of difficulties, collect incidents that can be transformed into training scenarios, and guide the feedback process to help trainees learn from their experiences.

One question we are often asked is about the relationship between cognitive skills training based on CTA and more traditional training approaches such as instructional systems design (ISD). ISD typically has been used for *procedural* tasks, so cognitive skills training could complement ISD by providing a *cognitive* perspective.

However, specifying the cognitive functions has the potential to make ISD more cumbersome and harder to use. Critics of ISD (e.g., Gordon and Zemke 2000) see the ISD approach as already being unwieldy, and inefficiently carving the world into too many small slices. Adding the cognitive dimension could make ISD even more so by overlaying all the procedural objectives with cognitive ones.

CTA efforts should result in richer training objectives, not longer lists of objectives. We do not see the decision requirements as a discrete set of training objectives. Rather, they are different facets of performing the cognitive work. Training developers would not want separate modules, say, for problem detection skills and sensemaking skills. The same scenarios should be training sensemaking and problem detection and decision making. The design process in cognitive training aims at providing experiences that permit trainees to explore, reflect, learn, work through confusion, and develop deeper and richer mental models for carrying out complex tasks.

13   Understanding How Consumers Make Decisions: Using Cognitive Task Analysis for Market Research

The decision making that surrounds the purchase and use of products is so pervasive and so common, it is almost invisible. Many of the products and services we buy are deeply embedded in our daily lives and are part of our daily habits and patterns of living. The cognitive activity that surrounds their purchase and use can seem mundane, or lacking altogether. How much mental activity is there in brushing your teeth each day, or taking a daily multivitamin, or doing the weekly grocery shopping? The answer, in many instances, is quite a lot. Cognitive Task Analysis (CTA) is useful for shedding light on a range of different decisions and cognitive questions, such as:

- How do consumers decide whether to purchase a product?
- How do consumers decide to change products after using one brand for months or years?
- What does it take to get consumers to consider a different product?
- How do consumers make sense of what the product does or how it works?
- What mental models do consumers adopt about how to use a product?
- When consumers have developed incorrect or inadequate mental models, what would help them shift to better models so that they are more satisfied with the product?
- What counts as product satisfaction?
- What does it mean to "trust" a product?

Traditional market research has not paid much attention to the cognition involved in everyday consumer activities. Certainly market researchers have spent time and money to understand the "purchase decision." But the path they have taken has had an overwhelming emphasis on the attitudes, beliefs, and preferences of consumers. What consumers know, how they think, and the strategies they have developed around the purchase and use of products are topics that have been largely overlooked by market researchers (Readinger, Ross, and Crandall 2004). Addressing these issues requires different questions, different tools, and a shift in focus from the typical market research approach.

In this chapter, we describe how CTA can be useful for understanding consumer choices and what we have learned about how to apply CTA methods in this arena. We begin with an example:

Example 13.1
The Bargain Hunters

Klein and Militello (2001) described a study performed for a commercial client who wanted to understand how some customers would ignore his product's quality and select the most inexpensive brands. An analogy would be purchasing a box of cereal, a five-second decision made while walking down the aisle of a supermarket. The client was losing this important segment of the market and was preparing to roll out a new economical version of the product. Before making this rollout decision, the client wanted to gain a better idea of the decision-making processes of these customers.

The client had been working on this problem for several years. The marketing group had performed a number of surveys and focus groups and had developed a decision model based on those data. This model provided a comparison of different types of customers and it had a slot for customers who were driven primarily by cost. But that was all the model said about these customers. Although the marketing group was satisfied with this model, they had been advised that a CTA might provide additional insights. The group only had a month left before bringing the research results to their upper management. They were not interested in a large-scale, comprehensive CTA. Instead, they just wanted to see whether they had missed anything important. They wanted a validity check on the data they had gathered, and the model they had so painstakingly developed.

The item was typically purchased at a supermarket, and collecting in-depth interview data at the point of purchase did not seem feasible. Instead, we decided to use a shopping simulation to get at consumer strategies and decision making. The project team prepared photographs that showed relevant sections of the supermarket aisle. The photographs were made into several panels, approximately four feet by two feet. The panels were set up on a table to simulate the display case. However, pricing information was not displayed in the photographs. Together, the set of panels depicted approximately thirty different product selections.

The client was interested in understanding a particular market segment, so the project team conducted a total of twelve interviews with consumers who represented that segment. The team consisted of two pairs of interviewers; each pair conducted individual interviews with consumers for approximately two hours. Working on consecutive days, we were able to complete data collection in only two days. In each interview, the participants were asked to imagine themselves walking through the supermarket and then turning into the aisle that contained this particular product. In this portion of the interview, we used a "think aloud" methodology. Participants were shown the display case photographs, and asked to look over them, as they normally would, and to describe what they were noticing and thinking. Because prices were not shown in the photos, participants had to request that information if and when they wanted it. Once the shopper had made her selection, the interviewers went back over what they had observed and queried the shopper about her goals, selection strategy, perceptions, and histories of use with various brands in the display. They also probed about her expectations for performance of the product she had chosen. After each set of interviews, the interview teams met with the client (several members of the marketing group were observing the interviews from behind a one-way mirror) and talked over what they had seen and heard. Much of the data analysis occurred interactively with the client, between interviews, and in discussion sessions at the end of each day of data collection.

At the end of the two days, the client understood that their model of "economical consumers" was too crude. There were decision strategies and nuances of choice that it simply was not capturing. All of the participants had clear preferences and dislikes, and none of them made the purchase decision purely based on cost. Their strategy was to identify three to four brands that they found acceptable in terms of performance and also that their families would be happy with and would use. Participants reported that they often looked through the newspaper before shopping to find coupons for cost reductions. In the supermarket, they continued to seek discounts, looking for store coupons and sales. When they got to the relevant aisle, they would quickly choose whichever brand from their "acceptable" set of three to four that was least expensive on that day.

The strategy enabled these participants to maximize a bargain. They could save money while still purchasing a brand their families would use. If their set of acceptable items had been smaller—say, fewer than three—then the chances of finding a significant discount were too low. If the set was too large, then the mental work of calculating and comparing cost became too great and took too much time. These economy-minded consumers had found an efficient strategy for balancing the factors that mattered to them.

At the end of the two-day data-collection effort, the marketing group realized that the CTA interviews helped them understand the "economical consumers" in a very different light. They later reported that the insights gained from the CTA interviews had contributed directly to one of the most successful product rollouts in their division's history.

Contrasts with Conventional Market Research Methods

The CTA approach to market research illustrated in example 13.1 is very different from typical approaches to studying consumers. Conventional market research is aimed at revealing the consumers' behaviors, attitudes, preferences, and beliefs. In contrast, CTA methods provide tools for studying the *way* consumers think about purchasing and using products and examines *how* consumers make decisions.

Referring back to example 13.1, we never asked the shoppers *how* they made their purchase decisions. Instead, we asked what they were noticing, which products they were considering, which products they were dismissing, whether they had tried any of the products they were now dismissing, and what their experience had been with various products. We *inferred* their decision strategy from what they told us, but we never asked them directly what their strategy was. Nisbett and Wilson (1977) demonstrated that people have a difficult time talking about their cognitive processes and cannot accurately report the way they make judgments and decisions. Nisbett and Wilson were actually conducting experiments about a very different sort of cognitive event—what happens when people try to articulate the basis for certain kinds of social judgments and comparisons. But their research offers an important insight about consumers and about people in general. It is unlikely that the shoppers interviewed in example 13.1— or any consumer population—could have told us their decision strategy even if we had asked them. Our method was to watch the strategy in action, to watch it over and over again across the different participants, and to ask questions about what we were seeing.

Market researchers use a wide array of research tools to understand consumers. Large-scale surveys, focus groups, one-on-one interviews, projective techniques, and observations of behavior are all part of the market researcher's repertoire. In example 13.1, the sponsor told us at the outset that the company had conducted interviews in the form of focus groups and had found these sessions informative. Even so, the sponsor was surprised when we said we were going to conduct a two-hour interview with each participant. "How can you spend two hours talking about what is basically a five-second decision?" they asked. They were accustomed to spending time talking with consumers, but they were skeptical that they would learn very much from a study of the consumers' decision making. Cognitive Task Analysis provided the client with a fresh perspective and a very different way of thinking about consumers.

It is not that market research is somehow missing the mark altogether. But there is a gap that CTA methods can fill, that can provide a significant advantage for market researchers. The ultimate goal of market research is to understand and influence consumer behavior. The prevailing notion seems to be that the best way to do that is by understanding (and influencing) consumers' attitudes, their preferences, and their beliefs. The role of most qualitative market research is to get consumers to report on those factors. By getting people to tell what they want and what they like, market researchers assume they can predict and influence what people actually do. This approach is not wrong; it is simply incomplete. It misses two key elements: (1) how consumers think about products and make decisions, and (2) the context that surrounds the purchase and use of products.

Value Added

In our work[1] with consumers, we have found CTA methods useful for shedding light on a wide range of issues surrounding the purchase and/or use of a diverse set of products and services. To date, our projects have focused on the purchase and use of adult nutritional supplements, laundry products, lawncare equipment, healthcare plans, prescription medications, photography tools and supplies, infant nutrition and care, home entertainment systems and services, pet care, and family meals. In each case, we have found that CTA illuminates aspects of how consumers think about products and their use. More often than not, sponsors are surprised at what they learn, in part because conventional market research has paid so little attention to how consumers think. Our findings tend to fall into three main types of results.

Consumer Decision Making

Cognitive Task Analysis methods can reveal the strategies consumers use to make the purchase decision. Cognitive Task Analysis tools are also helpful for illuminating the reasons people have for choosing or not choosing a particular product. They can provide insights about the functions consumers expect the product to serve, what it "does" for them. They can give sponsors a much deeper and more detailed understanding of consumers and the factors that drive choice and preference.

For example, for many consumer products, the common view is that the purchase decision occurs "in the moment" in the grocery or drug store aisle. An intriguing finding that has emerged in several of our studies is the importance of understanding deeper, longer term influences: family practices (e.g., "I was raised on Brand A toothpaste and I wouldn't use any other"); critical events (e.g., "I watched my grandmother become crippled with osteoporosis, and that's why I drink lots of milk"); patterns of product use (e.g., "I always pre-soak my husband's shirts for a day before I wash them..."); and trust (e.g., "I used product X for fifteen years and never used anything else. Then they changed the smell/look/ packaging, and I haven't used it since").

Consumer Mental Models

Mental models are internal representations of the external world. Cognitive Task Analysis methods provide a way to elicit and document these internal representations. They can provide a new perspective on how people think about the product and its function and what makes it effective or ineffective. Information about consumer mental models provides a deeper and more fully elaborated view of the meanings people have developed, the ways in which they understand a particular aspect of their world, and how a substance, product, or service may fit within it.

One aspect of mental models is the consumer's view of product design and use. Cognitive Task Analysis methods can provide a more thorough look at how the product is used and the context in which that use occurs. Understanding what people are doing, how they are doing it, and what they are trying to accomplish provides valuable information about necessary features, product placement, and product design.

Information Sources

How do consumers find out about a product or service? Where do they turn for information, what do they pay attention to, what do they discount, how do they use the information they receive? What is helpful, what is confusing, and what "turns them off"? Cognitive Task Analysis can provide information about critical avenues of influence and the pathways by which consumers arrive at a product or service. Our experience is that sponsors are often surprised by these data. Their assumptions about critical information sources are often incorrect or incomplete. In the section that follows, we describe methods and techniques for conducting CTA with consumers.

Studying the Way Consumers Think

Applying CTA methods to market research involves three basic strategies:

1. Observing consumers in the process of purchasing and/or using a product, and interviewing them during or immediately after the observations;

2. Creating a simulated task or purchase situation, or using props or other aids as the basis for data gathering; and

3. Interviewing consumers about product decisions they have made in the past.

The three strategies can be used separately or together, and work well when used in combination.

Concurrent Observations and Interviews

One of the challenges of doing interviews with consumers is that so many of their purchases and use patterns are embedded in everyday routines. On one hand, consumers may have significant experience, even expertise, at the tasks and routines that surround product purchase and use. On the other hand, it is not unusual for consumers to describe using a product while thinking about something altogether different. It can come as a surprise to sponsors to realize that consumers are simply not thinking as deeply about the product and its various features and opportunities as they themselves do. This aspect of consumer cognition makes its study quite different from much of the work described elsewhere in this book. Nurses and doctors, pilots, military commanders, soldiers, and firefighters all have routines they follow. However, the risks and responsibilities associated with their daily work require that they manage their attention. They have to maintain focus and concentration on what they are doing. This simply isn't the case for many kinds of product use. It is perfectly safe and reasonable to think about the day at work while loading up the dishwasher. For these reasons, we have found that opportunities to observe consumers in interaction with the product, and to interview them in concert with those observations, are often essential.

Finding ways to watch consumers make decisions while purchasing items in stores or other commercial settings can get tricky. Store managers are understandably reluctant to let researchers loiter in the aisles with clipboards, tracking the actions of consumers. If they appreciate how the findings will be used, and if they see some direct benefits from the findings, store managers are more likely to cooperate. Anthropologists (e.g., Underhill 1999) have made great use of observational methods to study consumer behavior. However, a CTA project requires that researchers have an opportunity to communicate directly with the consumer. Pure observation without any form of interview or any chance to probe what people are thinking about is not CTA. By coupling direct observation with concurrent or subsequent interviews, the CTA probes become very effective—the interviewers are able to ask about details of actions and events they have observed, not about remembered events or hypothetical cases.

Studying the ways consumers *use* a product has fewer complications. Many companies conduct qualitative market research by going to the homes of consumers and watching them as they do tasks and chores that require a particular kind of product. The context in which a product is used can be critical to understanding purchase decisions. During an interview, researchers might never think to ask about where the product is stored, or how it gets from the store to the basement. However, watching a sixty-year old woman climb a stepladder to reach her storage shelf while balancing a container of detergent, it becomes very clear why she never buys larger, more economical packages. Notice that the goal of these qualitative studies is to make discoveries about the way consumers carry out their tasks. In CTA market research, the goal is to link observations of product use with probes about what people are thinking and noticing, what they are trying to achieve, what they expect the product to do and not do, what they consider success and failure, and what they have tried in the past that has worked or not worked. Answers to these probes can provide additional important insights about why consumers use products in particular ways and how they think about what they are doing.

Simulations and Props

Sometimes it isn't possible to observe consumers in action. Sometimes we can learn more about decision making by setting up a good simulation. Example 13.1 illustrated how a simple simulation, consisting of high-quality, life-size photographs of a grocery aisle, provided an effective platform for probing consumer cognition—seeing what they were noticing, what they were considering, what they were ignoring, and what they were analyzing. In other situations we have used props, prototypes of products, or graphical representations and asked consumers to tell us about the product, or to show us what they do with it, or how they use it. The photograph, prop, or prototype provides a basis for both observation and interview data gathering.

Example 13.2 illustrates the use of graphics to explore consumer mental models of how a product works:

Example 13.2
Dirty Secrets

The client was very frustrated with its customers. The client's product, a form of detergent, was very effective for removing stains from clothes. But some customers didn't realize this, and didn't believe it despite an expensive advertising campaign to convince them of its efficacy. The product contained a set of powerful enzymes that removed most stains and cost much less than specialty stain removers. The project manager was frustrated with his inability to get a key brand concept across to consumers. Even people who were loyal to the brand and had used it for many years seemed to have no understanding of how it worked. The project manager wondered how they could be so dense.

It was harder than we thought it would be to figure out what purchasers of this product were thinking about. At first, we tried the direct approach in our interviews. One of us (G.K.) was dumb enough to ask a participant, "So, what is your mental model of the way that stain removal products work?" The participant turned to the other members of our interview team with an incredulous look on her face, as if to say, "What is this guy talking about?" There was a lot of unpleasant giggling in the car as we drove back to the office afterward.

So if we weren't going to get anywhere asking people how stain removal products worked, how were we going to get at this? We tried various ways of rephrasing the question: "How do you think about stains?" "How do stain removers work?" The consumers we interviewed continued to give us blank looks. They were doing their best to be helpful, but they appeared to be truly baffled by our questions. They said they never gave these issues any thought. They didn't know the chemistry involved, didn't know anything about enzymes or how enzymes might function; they just knew what worked (at least based on their experiences).

In desperation, we came up with a different CTA strategy. We had heard some snippets in the interviews—occasional phrases about "sliding the dirt out" or "bonding to the dirt and lifting it out" or other fragmentary images. We took these images and created a set of seven different one-panel cartoons, each illustrating a different potential mechanism that the stain removal products might be employing (figure 13.1 contains examples of the cartoons used in these interviews).

Then we went out for some new interviews. This time we showed product users the cartoons and asked which was the closest to the way they thought the stain removers worked. And if the cartoon they chose wasn't exactly right, how should we change it? Paydirt. Our interviewees enjoyed studying each of the cartoons and telling us what was wrong with them, and what to do to make the pictures "right." Finally, we were able to capture their mental models, as long as we never used those words.

Retrospective Interviews

The core of CTA is the interview, whether it is coupled with observation or with simulations. Do retrospective interview materials make sense in this context? Often, they do not. Think back to example 13.1. A consumer might take five seconds to reach for a box of cereal while walking down the supermarket aisle. The retrospective

Figure 13.1
Cartoon examples of stain-removal mechanisms.

incident-based methods described in chapter 5 didn't make much sense in this case. What would the interview cover? What sort of timeline could we construct for such a fleeting event? Similarly, in example 13.2, we found that we needed to augment the interview with a different strategy.

However, we have used retrospective interviews successfully in market research CTA. One example was a project to understand how consumers think about brand loyalty and what prompts consumers to switch brands. The sponsor wanted to understand the conditions that would make a customer give up a favorite brand and turn to another. In this case, there was nothing to observe or to simulate. Instead, we used critical decision method (CDM) interviews to examine instances in which the customer had decided to shift brands. In each interview, we identified the consumer's history with different brands—the one she started using while living at home, what she used when she first moved out of her parents' home, and each subsequent shift. We treated every brand shift as a decision point and focused the interviews on the factors that contributed to the decisions to change brands. After interviewing about twenty consumers, we were able to provide the sponsor with a model of what it takes for consumers to change their loyalties.

An approach to CTA interviewing that is designed specifically for market research is called the Knowledge Audit for Market Research Applications (KAMRA), blending aspects of the CDM and Knowledge Audit methods. It is designed to elicit incidents of a particular type, for example, incidents of purchase, incidents of use, and incidents involving surprise about an aspect of product performance. Timelines are obtained, when appropriate, and then specific probes derived from the Knowledge Audit are used to explore particular aspects of expertise and decision making. The KAMRA technique allows for very focused probing on lived experiences that have high information value, given the goals of the project. KAMRA has been particularly useful for understanding mental models and for uncovering aspects of expertise and expert-novice contrasts (e.g., Readinger 2004).

Cognitive Task Analysis interviews let us understand how consumers view their tasks and their problems. That information can be important in terms of positioning a product. It can also reveal information that is key to developing new product concepts and to product design. By appreciating the client's perspective, product development teams can better understand why certain features matter and why others are rejected. The next example illustrates the disconnect that can exist between the way the product developers think about the task and the way the consumers think about it.

Example 13.3
Channel Surfing Safari

The client was developing an ambitious system to provide television viewers with ready access to hundreds and hundreds of channels. However, the client feared that people would be overwhelmed by having so many choices. Therefore, the product development engineers generated a comprehensive format for listing all the available shows, and for prioritizing the list in various ways—to emphasize situation comedies, movies, game shows, and the like. This customized menu was the entry into the system. The customer could access this guide at any point to view all the television-viewing choices. At the beginning of a serious evening of television watching the customer could review the options for the rest of the evening, make some selections, and get started with the choice for that time slot.

We conducted a series of observations combined with interviews and discovered a serious disconnect between the television viewers and the product design engineers. The design engineers resonated to the scheme of scanning the options for each half-hour so that they wouldn't miss out on exciting possibilities. The designers liked the idea of being able to sort in different ways, create lists of favorite channels, and conduct all sorts of clever manipulations of the guide. They disliked the process of channel surfing, with all of its randomness and uncertainty, and found it frustrating.

The television viewers we studied had a very different perspective. They didn't like the idea of scrolling through a spreadsheet of shows. For them, it was a waste of time. When they turned on a television, they wanted to start watching shows. They didn't mind channel surfing. In fact, they would rather surf and see what caught their attention than spend time scanning a list in order to make a selection. They wanted to be watching television, not making choices and preparing to watch television.

Using CTA with a Sales Staff

Many purchases are made through a salesperson, instead of directly from a store shelf or Internet site. Cognitive Task Analysis provides leverage for understanding how a sales staff views its customers. A company may want to ensure that its sales force knows how to size consumers up and make use of that appraisal. It may want to evaluate the sales staff's mental models of the consumer and the consumer's view of the product and its features. It may want to understand what distinguishes its best salespeople from the rest of the staff. The following example describes a project for which we interviewed both the sales staff and the company's customers. The company used our findings to implement a training program for their new hires. They also revised their models of this consumer market, including segmentation and marketing strategies, based on findings from the CTA.

Example 13.4
A Double-Barreled CTA

A successful consumer product manufacturer, Company B, was in second place in terms of its market share, and couldn't find a way to catch up to the leader. Company B had a high-quality product offering that had been on the market for a decade. Their product was of similar quality and price to the market leader, and was targeted at the same segments. The problem was that Company B had entered that market many years after its competition.

Recently, funding had become available to allow their sales force to grow, and Company B wanted to take full advantage of this opportunity. They already employed a collection of experienced, highly skilled salespeople. They wanted to know how to make their less-skilled salespeople great, how to make their great salespeople even better, and how to give their new hires a training program that would help them succeed as quickly as possible. To determine this, we needed to understand two things: How did the market think about the products that were being sold and the techniques for selling them, and what separated Company B's high-quality salespeople from their mediocre ones?

We arranged sixteen interviews with consumers from locations at several sites across the country and conducted interviews and observation aimed at understanding how they used the product and how they made the decision to purchase this product. We identified different categories of consumers that reflected their understanding and interest in the product, and the way they thought about themselves as consumers. Both dimensions were important in determining consumers' buying behavior, as well as the style of salesmanship that they preferred and found most effective.

Next, we tried to understand the thinking and strategies of the sales force. We interviewed twelve salespeople using the KAMRA interview method described earlier in this chapter. We learned that the difference between excellent salespeople and their struggling counterparts was really a matter of degree. In fact, of the twelve salespeople we interviewed, virtually all were able to discuss the mental models of their customers in some detail; their understanding of this aspect of the sale was well developed and sophisticated.

The advantage of expertise was primarily in the sheer number of strategies that the highly skilled salespeople had for dealing with clients who held different mental models. Whereas a novice salesperson was able to identify the "type" of customer he or she was dealing with, the expert salesperson could identify the type, and then select from a rich set of strategies that were appropriate for this customer. When one strategy wasn't working, there were several possible backup plans available. The less successful salespeople simply lacked this variety and depth of response. Although they had learned the critical cues to identifying and understanding their market, they struggled with developing back-up strategies when the initial one failed.

The sponsor for this research used the findings to develop a training program for newly hired salespeople and also to provide "refresher" training to experienced salespeople. The training emphasized the critical cues and the segmentation of the market we had identified as key elements of salespeople's perceptions. The strategies of the most successful salespeople were also documented and used during training to provide suggestions and alternatives to those with less experience. Finally, the company altered the structure of its sales force in order to encourage this type of transfer of knowledge and skill in the future. Here, the goal was to disseminate insights as the new sales force developed more sophisticated strategies and a deeper understanding of the market and client base.

By understanding how their consumers were thinking about their product, Company B could better appreciate the expertise that their salespeople possessed. In turn, the sales force was able to use this experience to their advantage and train a new generation of salespeople in the best practices of the organization. The understanding they gained of their consumers' thinking essentially helped them begin to gain on the leader.

Summary

Market researchers are primarily interested in figuring out how to influence consumer behavior. Conventional approaches to market research have pursued that goal by studying consumer behavior, attitudes, desires, and beliefs. Although all this information is valuable and important, it does not address the critical component of consumer cognition: what consumers know, how they think, and what strategies they have developed for buying and using products. This chapter explored ways in which CTA methods can be applied to market research questions. Cognitive Task Analysis methods can reveal the strategies consumers use to make the purchase decision—whether to purchase a particular product, which brand to select, which features to choose, or whether to try a different brand with different features altogether. Cognitive Task Analysis methods can explain how users understand what products do and how they work, and can offer insights about why consumers may not use products in the ways that their developers intend. Finally, CTA methods can reveal the skills and the gaps in understanding that front-line sales staff may have regarding their customers.

14 Cognitive Task Analysis for Measurement and Evaluation[1]

Measurement and Cognitive Task Analysis (CTA) are related in a number of ways. For example, throughout this volume we have discussed methods and methodology in reference to domain practitioners and experts. Clearly, an important issue is how to identify experts or individuals at other levels of proficiency. Thus, measurement can be involved in identifying research participants.

Proficiency levels can be scaled in a number of ways. Career interviews can detail individuals' training and experience. Individuals with specialized knowledge and skill can be identified through sociogrammetry—asking practitioners about where they go for advice (Stein 1992, 1997). In some domains, proficiency can be evaluated by looking at actual performance measures. If CTA research is predicated on any notion of expertise (e.g., a need to build a new decision support system that will help experts but can also be used to teach apprentices), then it is necessary to have some sort of empirical anchor on what it really means for a person to be an "expert," say, or a "journeyman," or "apprentice." Ideally, a proficiency scale will be both domain- and organizationally-appropriate. For example, in the weather forecasting project (Hoffman, Coffey, and Ford 2000) it was necessary to refer to senior and junior ranks within the expert and journeyman categories. Ideally, a proficiency scale will be based on converging evidence from more than one scaling method. A detailed discussion and illustration of a number of proficiency scaling methods appears in Hoffman, Trafton, and Roebber (2006).

In this chapter we focus on two of the main intersections of CTA and measurement. We begin by showing how CTA can be used in measuring performance. Next, we describe ways of measuring CTA methods themselves.

Measuring Practitioner Performance

Cognitive Task Analysis methods can be particularly useful in assessing forms of cognitive support, including cognitive training programs and decision support systems.

These programs and systems are designed to help people make better decisions, understand the significance of events, and do a better job of planning and performing other macrocognitive functions. If we want to gauge effectiveness we have to examine the impact of the programs and systems on the way people think. This is where CTA comes in.

Measures can be created to capture important characteristics of the nature of performance and change. The resulting measurements can be put to many uses, such as setting goals and tripwires, regulating performance, ensuring compliance, and promoting fairness (Klein 2004, chapter 14). As an integral part of iterative design, measurements of performance can help CTA researchers see if a system or program addresses a problem situation fully enough. Measurements can also be brought to bear on competing concepts, to help select a superior approach, or to consider whether a particular design is holding up under changing conditions. Ideally, measures are not used just once in a project (say, to evaluate performance after a new software system has been delivered to the sponsor), but are used repeatedly in system development. In this way, measures provide both baselines and checks and balances to intuitions about the nature of performance and changes.

In the field of cognitive systems engineering (CSE), the phrase *cognitive measures* or *cognitive metrics* is used to refer to indicators that tell us something about the state or process of cognitive work. Cognitive measures are the rulers by which proficient performance (human and/or system) is gauged. Is a practitioner improving performance over time? Is the teamwork getting faster and increasing in accuracy? Are decision makers responding promptly to changes in the situation? Are practitioners more regularly accomplishing their mission or meeting the schedule? Do people recognize more problems, or recognize them earlier, or do they increasingly recognize subtle ones and their potential trajectories?

When we speak of cognitive measures, we are basically talking about enabling comparisons that help us as researchers better understand some phenomenon, state, or process and how it may be either stable or changing. Key to the proper use of cognitive measures is the concept of comparison. We can compare two or more measurements, or compare measurements against some standard, estimation, or judgment. We can compare groups (e.g., novices and experts, users and non-users of systems), we can conduct pre- and post-tests, we can compare expectations to performance, we can track trends over time to seek dependencies on selected variables, and so on.

However, these kinds of comparative evaluations are regrettably rare. Too often, the evaluation of new systems is a weak add-on to system development efforts. In many information technology (IT) system development efforts, evaluation is typically based on a "satisficing" criterion—that is, users work with the new system for a while and then are queried concerning their opinions, resulting in evidence that some of the people like it, more or less, at least some of the time. Sometimes, the only evaluation is whether a senior member of the organization likes the new system or training program, after perhaps a brief glimpse or PowerPoint presentation of design features. Typically, when users are queried about their reactions to new systems, the sole metric is vague user satisfaction, "So, tell me, how do you like it?" In the evaluation of new systems, global satisfaction ratings are simply not enough. All they show is that some people said that they like the new system or training program some of the time. They do not inform the researchers about ways that the new system might make old tasks more difficult. They do not inform about how the fundamental nature of the job might change. Most often, satisfaction ratings result in some suggestions for minor changes in the interface, but no fundamental evaluation or improvement.

Clearly, this is not sufficient. We can tap into the large and growing literature on CSE to see pointers to a great many things that can and sometimes should be the subject of metrical evaluation. For instance, when a new technology is introduced there should be:

Gains

Usefulness

Usability

Effectiveness/success

Enhanced immersion ("being in the problem")

Enhanced direct perception, recognition, comprehension

Accelerated achievement of proficiency

Enhanced intrinsic motivation

Enhanced capability to cope with rare or tough cases

Enhanced capability to recover from error

Adaptability or resilience

Reductions

The gap between the actual work and the true work[2]

Mental workload

Time and effort

Negative affect such as frustration

Overconfidence in the technology

Excessive mistrust in the technology

Avoidances

Working the technology ("make work")

Fighting the technology ("workarounds")

Misunderstanding the technology ("automation surprises")

To date, there have been few attempts to measure most of these. We hope that in the future all the macrocognitive functions described in chapter 8 will have associated measures. Mica Endsley has paved the way here. She developed and evaluated measures of situation awareness, including the Situation Awareness Global Assessment Technique and the Situational Awareness Rating Technique (Endsley and Garland 2000; Endsley et al. 1998). None of the other macrocognitive functions have comparable measures with a proven track record of utility across domains, although such work is underway. Of course, in natural settings the various macrocognitive functions are tied together, and having separate measures for each may not be sufficient.

Process and Outcome Evaluation

When we design a new system or prepare a new training program we need to know if it is any good. We can assess the effect on performance, looking at the outcomes of the training or the technology. Can people do the work more quickly and/or more accurately? We can also assess the effect on the way people do their work—the processes they use. Are they applying better strategies, or are they aware of more features of the situation, or do they have better mental models? All of these might reduce their mental workload.

Outcome evaluations describe the impact of an intervention, and *process evaluations* are intended to reveal the reasons for this impact. The difference is between what is happening and why it is happening. System developers are usually interested in both outcomes and processes.

Measuring outcome is critical. The developers (of an application of IT or of a training program) want their system to have an impact on the way people do their work. The developers and sponsors may rely on standard outcome measures such as time, speed, accuracy, and costs. They can also look for indicators of outstanding performance. What is it that distinguishes people in the eyes of their peers? What makes for a high-quality product? What makes people good to collaborate with, and what makes for a useful collaborative session? What would make a task easier? Knowing what experts can do that novices can't often provides ideas for outcome measures.

Feedback about process is often more helpful than feedback about outcomes. Process evaluations are particularly useful in training. Process feedback shows the learner how to make changes that will improve performance. It shows the instructor where there is good acquisition of critical cognitive skills and where there are gaps. In the same way, evaluations of process can allow the CTA researcher to learn. Process metrics help system developers and sponsors understand why performance levels are disappointing and how to improve them. For example, if a design team did a poor job, the IT application they produced may lead to performance that is worse in some respects rather than better. The design team would need to study the processes used in order to appreciate what is going wrong that the system developers did not anticipate (e.g., how the system adds more mental workload by causing a need for workarounds which interfere with the job).

Let us use the illustration of a consumer who wants to purchase a powerful automobile—one that provides ready acceleration. The consumer, in considering a certain model, might inquire into the performance, such as the time needed to go from zero to sixty miles per hour. The consumer can also study the processes that affect performance, such as the horsepower or torque of the engine. The consumer can compare different models on these outcome (time from zero to sixty) and process (horsepower) measures.

Notice that acceleration speed and horsepower are both quantitative data measures. As we discussed earlier, quantitative measures sometimes do not tell the entire story. It is the rare circumstance, especially in the design of information technology, that a single metric will tell the entire story. Quantitative measures of outcome can be misleading—a car that can go from zero to sixty in less than six seconds may be impossible to really "drive" if the owners spend most of their time in congested traffic. Quantitative measures of process can also be misleading. An engine with 140 horsepower might do well in a subcompact, but would be insufficient for a pickup truck.

The overall point is that CTA needs to rely on both quantitative measures and qualitative forms of assessment, and we need to evaluate both outcomes and processes. The following example from a study with Army information operations officers (Klein et al. 2002) illustrates how nonstandard metrics of process helped us interpret a standard metric of time to completion.

What's Behind the Numbers?

We were conducting a study of sensemaking (Klein et al. 2002). We set up a simulation study in which we presented a series of situation reports to experienced and apprentice Army Intelligence officers. Data were collected on the number and types of implications, inferences, speculations, and explanations the participants developed as they read the stream of messages. These were coded and counted. We found that the experienced participants, the experts, made more inferences than the younger officers. Their experience, their richer mental models, made it easier for them to generate inferences. Six apprentices went through ten scenarios—almost two scenarios per apprentice in the two-hour time block. In contrast, the six experts only went through seven scenarios—only one of the experts was able to complete a second scenario in the two-hour time block—because the experts were taking so much more time reporting inferences.

Thus, while "time to completion"—a standard metric used in evaluations of human performance—may seem relatively straightforward, the *nature* of cognitive activity during the time to completion was more compelling for understanding performance. The important discovery we made using CTA methods—that richer mental models enable experts to generate a greater number of inferences—would have escaped us if we only used the outcome metric of time to completion.

Attempts to quantify process measures can also be misleading if the measurements are taken out of context. As an illustration, Dekker (2003) has lamented the use of simple metrics such as frequency counts in categorizing the reasons for aviation mishaps. The frequency counts (e.g., number of mishaps attributed to loss of situation awareness) do not do justice to the causal relationships that trigger accidents in complex human-machine systems; accidents are best represented as rich incident accounts. Such "stories" (that is, qualitative data) help us appreciate and understand the metrics. At the same time, the metrics help us clarify the trends we might sense in reviewing the qualitative data.

The line between outcome and process evaluations is sometimes blurred. At any given level of scrutiny, the performance being studied is the outcome and the contributing factors are the process. But the outcome at one level can be a reflection of the process for the higher level, or vice versa. Building an engine with greater horsepower is a process by which we achieve faster acceleration, which is a process by which we move through traffic more quickly, which is a process for reducing the time we spend commuting. Achieving faster and better diagnoses in an emergency room is a process for improving patient care, and is an outcome of having medical staff members with better mental models, which is an outcome of providing staff with better training and equipment, and so forth.

Nevertheless, at any given level of inquiry we can distinguish between the outcomes of interest and the processes that govern or influence these outcomes. We can distinguish between quantitative measures that ground our conclusions and qualitative considerations that provide the context for understanding these measures.

Cognitive Task Analysis applies to each of these forms, quantitative and qualitative, outcome and process. The team designing a new decision support system needs to evaluate whether or not it improves decision making. Cognitive Task Analysis studies may show that more experienced commanders can make faster and better decisions. So the designers may try to compare decision speed (quantitative) and ratings of decision quality (quantitative/qualitative) with and without the new system. Designers may compare the rationale for decisions (qualitative) with and without the new system, using CTA methods to see if the new system helps the operators achieve insights. Decision speed and quality are also outcome measures for such projects, whereas decision rationale describes process.

Similarly, training developers may try to measure the effect of their training programs or simulators. They may look at the impact of their work on performance. They may also measure the impact on the processes underlying performance. Cognitive Task Analysis studies can capture expert-novice differences, and instructors can convert these differences into measures—progress markers—to help students assess how rapidly they are improving. Cognitive Task Analysis studies can identify barriers to achieving proficiency, suggesting qualitative evaluations that instructors can apply to diagnose the reasons why trainees are still struggling. Cognitive Task Analysis methods can be incorporated into debriefs after exercises to reveal how trainees made decisions and achieved, or failed to achieve, situation awareness.

Cognitive evaluation is an activity to help researchers see the link between cognitive processes and performance. Cognitive evaluation highlights the points of comparison that reflect some cognitive state or process and how it is changing. Cognitive Task Analysis can help researchers and system developers in developing, selecting, conducting, and interpreting cognitive evaluations.

Uses of Measurements: Identifying the Cognitive Requirements to Study and Measure

Cognitive Task Analysis can play a role in helping researchers to configure measures. CTA data highlight critical cues, patterns, and relationships, and can show what makes decisions and judgments so difficult.

Cognitive requirements are the challenging aspects of a task, along with the factors that are responsible for the challenges—the reasons for the difficulty, the types of errors people make, the types of strategies people need to employ. Thus, to appreciate the cognitive requirements for a nurse in diagnosing sepsis conditions in newborns, macrocognitive functions such as sensemaking and problem detection might be critical. Having an extensive knowledge base might also be critical. The CTA study performed by Crandall and Getchell-Reiter (1993) also identified a range of perceptual cues and patterns that skilled nurses can recognize. The researchers drew on that CTA methodology to identify the reasons why nurses sometimes have difficulty getting rapid treatment for septic infants. Cognitive Task Analysis data can also demonstrate what kinds of errors people make. In these ways, we believe that CTA findings can suggest context-specific measures of performance.

Thus, researchers and practitioners can use CTA to determine which aspects of cognitive performance they should be monitoring. Sponsors can use CTA to identify the aspects of cognitive performance they want to see improved. For example, cognitive requirements are often used to justify the creation of new systems. In many cases the rationale for a new application of information technology or a new training program will refer to some performance problem or series of accidents or inadequacy of current levels of performance. But what led to those breakdowns and inadequacies? Cognitive Task Analysis studies can help the sponsors identify the cognitive requirements of the work, particularly the cognitive requirements that are difficult to accomplish. Frequently, the rationale for a new project will refer to macrocognitive functionality: better decision making, better situation awareness, better planning, better adaptability, and better coordination. Our understanding of macrocognitive functions and the supporting processes (see chapter 8) can help guide the CTA effort to clarify these types of cognitive requirements so that evaluators have a more specific basis for their observations and measurements.

Cognitive requirements can serve as criteria for designing the project and implementing the system. Thus, they are part of a front-end analysis. The cognitive requirements can also be included in the final acceptance criteria for the system. Here, researchers are faced with the challenge of determining if the cognitive requirements are met. For example, the new decision support system or training simulator may need to boost decision making in urban combat settings. The validation can center on decisions about prioritizing threats, selecting courses of action for clearing a building, orchestrating a multiblock campaign, and so on.

The cognitive requirements of the task are both the macrocognitive challenges that the task poses (e.g., decision making, replanning, problem detection) and the requirements for addressing these challenges (e.g., building mental models, managing uncertainty, managing attention). All of these can point toward potentially useful measures. They can describe the aspects of cognitive performance that need to be supported and how (e.g., how much, what kinds, when, and for how long). They provide criteria against which to test and evaluate performance.

Table 14.1 is an example from a project that used CTA to identify metrics for systems that were being proposed to support intelligence analysts.[3] As Malek et al. (2004) performed the CTA study, they found that expert analysts know how and when to use a

Table 14.1

Metrics for supporting intelligence analysts

Cognitive Requirement	Measure	Data Collected	Expectations (if a decision aid is actually helpful)
Quality of sensemaking	Correctness and accuracy in projecting the future	Verification that forecast events occurred	Increase
	Comprehensiveness of the knowledge used in the sensemaking	Comparison of the propositions in the analysis with those in some gold standard, such as an analysis by a senior analyst	Increase
Quality of mental modeling	Level of validation of models/arguments/theories	Percentage of sources and data properly vetted	Increase
	Types of sources	Changes in the category system used, based on observations and probes	More appropriate
Decision making	Ability to manipulate disparate data	Variety of databases used	Increase
	Familiarity with a variety of information sources	Number of databases used	Increase
	Priority given to soft targets	Proportion of hard vs. soft targets considered, using activity logs	Increase
	Skepticism about data and collection procedures	Number of queries about data sources	Increase
Coordination and maintaining common ground	Effort needed	Number of messages, time spent on teamwork	Decrease
	Effectiveness of pass-ons	Time spent synchronizing knowledge bases	Increase
	Availability of and visibility to back-channel information exchange	Number of messages outside the formal chain of coordination, using activity logs and observations	Increase

variety of databases. The skilled analysts knew how to access and manipulate the data within the databases. They knew the important characteristics of the data (e.g., credibility, age) and when it was important to compare data.

The CTA data suggested a number of new measures. For example, one analyst suggested that the overall quality of arguments would be increased if analysts took into consideration data on "soft targets." Soft targets might include the beliefs of the population, rather than "hard targets" such as adversary command posts. The CTA revealed that analysts share important data in "back channel" exchanges (i.e., discussions that take place outside of formal channels of communication). These kinds of findings suggested the types of measures shown in column two of table 14.1.

Uses of Measurements: Designing Scenarios

In some cases, measures are built into training scenarios. Scenarios are the bedrock upon which evaluations are carried out. Without the proper scenarios, measurements can be useless. Cognitive Task Analysis methods help in rigorous and systematic creation of scenario suites that provide proper coverage and hence allow good judgments about the true nature of the performance of the system in use. Instead of asking for ratings or recall scores, researchers may "seed" scenarios with decisions or ambiguous situations or other cognitive challenges and then study how the participants respond.

For example, Baxter et al. (2004) assessed the ability of Marine lieutenants to master the cognitive requirements of an ambush scenario. The research team constructed and administered a variety of comparable ambush scenarios. The researchers also embedded periodic check-in radio calls between the lieutenants and their company commander. These calls showed how the lieutenants were interpreting the situation. The researchers also noted the actions taken by the lieutenants as evidence of decisions about how to respond. For instance, a message about enemy presence on a nearby ridge had implications for the mission. The researchers measured whether or not the platoon leaders notified the company commanders, to gauge whether the platoon leaders realized the implications. These data were compared prior to and after the training to demonstrate a significant improvement in performance.

Cognitive Task Analysis studies have often been used to design scenarios and to highlight the cognitive challenges that need to be incorporated into the scenarios. Phillips et al. (2001) described how to use CTA findings in constructing scenarios around the cognitive requirements of the task of clearing a building during urban combat (see chapter 12). Scenarios such as these can be used to measure the rate of learning for the trainees in terms of the number of scenarios they have successfully accomplished.

Uses of Measurements: Estimating the Impact of Innovation

Researchers can conduct interviews to determine if an intervention has achieved its objectives. The interviews can assess cognitive processes before and after or with versus without the new technology or tool. The interviews can include having the system users complete some rating scales. The interviews can elicit factual information in evaluating what participants have learned. CTA interviews can examine the types of mental models that participants used prior to training and after a training intervention. If the intervention was effective, the post-training interviews should show more accurate and sophisticated mental models. Researchers can insert CTA probes as participants are going through a scenario to elicit the types of cues the participants notice.

Project sponsors need to weigh the expected benefits of a new effort against its costs. When the program is a new decision support system or training simulation, the benefits will involve better decision making and other cognitive functions. For this function, CTA can contribute by helping sponsors identify the dimensions—the cognitive measures—that might best reflect the value of their investment. Cognitive Task Analysis can also help sponsors speculate about key questions such as: Is it possible? How likely or risky is it? How do we know we're on track? What kinds of changes will the investment will result in?

In table 14.1, the fourth column captures expectations about the desired impact of the technology insertion. Most of the expectations listed in this table are simply decreases or increases. Some sponsors will want more concrete data than are described in table 14.1. Others will be grateful to capture even a fraction of what is called out in such a table, realizing that the opportunity for thorough or extensive quantification might be limited by practical or funding considerations. But generally, sponsors do need some quantified measures or at least predictions about ranges, variability, and causal factors.

One type of impact to consider is anticipating unintended consequences. Information technology doesn't always improve operator performance. Information technology sometimes improves performance in ways that were hoped and anticipated, but causes new types of errors or degrades performance in ways that were not anticipated (Woods 2002). Sometimes the technology can get in the way. If we only consider metrics sensitive to the gains, we can miss the downside of the proposed system. Cognitive Task Analysis studies are useful (and sometimes necessary) for describing the cognitive requirements for effective performance, regardless of the technology. Advocates for the system will usually emphasize cognitive measures that reflect how well the system does its job. Too often the individuals who promote new technologies express their expectations in cognitive language, such as improvements to situation awareness and decision making—but then capture performance measures for the system without further consideration of the human users. Advocates for the domain practitioners need to draw on the CTA findings to include metrics that are sensitive to the ways the system might interfere with the exercise of expertise.

Pliske et al. (1997) provide an illustration. They studied skilled weather forecasters. Among their findings, they noted that experts focused on "messy" areas in the weather charts, areas of instability that the experts needed to watch more closely. Some proposed systems that purported to help the forecasters included techniques for smoothing the data to eliminate the unsightly messiness. Pliske et al. noted that such smoothing would actually make the job of the skilled forecaster more difficult. If such systems were delivered, a test and evaluation criterion could have been to measure the time the forecasters needed to notice the problem of the day, using the new system. That measurement criterion anticipates negative impact. However, the time to notice the problem of the day, by itself, says nothing about the nature of the potential negative impact. It would be more valuable to understand how the smoothing masked forecasting data. The story behind the data can be as important as the data.

Pliske et al. also documented how skilled weather forecasters needed to form their own mental models of weather patterns before consulting the outputs of the computerized weather forecasts. Some proposed decision support systems were designed to automatically provide forecasters with computer models of the weather situation. Depending on how and when they were presented, these automatic diagnoses interfered with the desire of the skilled forecasters to formulate their own understanding first. A test and evaluation metric in this case could have been the time needed to realize that the computer-based forecast was wrong. If the forecasters stopped building their own mental models and just relied on the system, they should take longer to discover instances in which the system was inaccurate.

Measurement to Answer Questions About CTA Methodology Itself

Cognitive Task Analysis itself will sometimes have to be subjected to evaluation based on measurement. How do we know that a CTA method is the best to use in a given situation? How do we know that the CTA was effective and productive? In many applications, if the CTA works, that is satisfactory. But in other situations, systematic evaluation of CTA is important or even critical.

An example comes from investigations that have attempted to compare alternative methods for eliciting the knowledge of experts (Hoffman 1987; Hoffman, Coffey, and Ford 2000). In the earliest days of expert systems, computer scientists relied on unstructured interviews (see Cullen and Bryman 1988). This led to what was called the "knowledge acquisition bottleneck." It took longer to interview the experts to reveal their knowledge and decision rules than it took to program the expert systems.

This issue encouraged a consideration of methods from psychology that might be brought to bear to widen the bottleneck, including methods of structured interviewing (Gordon and Gill 1997) and methods adopted from the psychology laboratory, such as "think aloud problem solving" (see Chi, Feltovich, and Glaser 1981; Ericsson and Simon 1993). In addition, expert performance could be studied by withholding certain information about the case at hand or by manipulating the way the information is processed (see chapter 6). In the "method of tough cases" the expert is asked to work on a test case (perhaps gleaned from archives), the idea being that tough cases might reveal subtle aspects of expert reasoning or some particular subdomain or highly specialized knowledge, or certain aspects of experts' metacognitive skills, for example, the ability to reason about their own reasoning or create new procedures or conceptual categories on the fly.

In considering the manner of comparison, Hoffman (1987) proposed that knowledge elicitation methods might be compared on a number of variables, including the simplicity of the task and materials, the brevity of the task, the adaptability of the task (to variations in the materials, instructions, participants, and so forth), the artificiality of the task (relative to the expert's familiar task), the validity of the data (i.e., correct information about expert knowledge or reasoning), and finally, the efficiency of the method.

Hoffman's concern was with creating a method for gauging the relative efficiency of knowledge elicitation methods. This involved the following considerations. First, in most knowledge elicitation procedures (as in most CTA procedures), the research begins with framing procedures and bootstrapping such as documentation analysis. Much expert knowledge can be culled from that, forming what Hoffman called a "first-pass knowledge base." Knowledge that would be revealed in elicitation procedures conducted subsequent to the documentation analysis could be regarded as "informative" if that knowledge is not already in that first-pass knowledge base.

The second factor in Hoffman's scheme involved a consideration of time and effort. Rather than looking just at the time it takes to conduct knowledge elicitation, Hoffman measured "total task time." This includes the time needed to prepare to run the procedure, to run the procedure, and to format the results in accordance with the mediating representation used for the knowledge base (some useful product from the knowledge elicitation).

To compare alternative knowledge elicitation methods, Hoffman (1987) calculated the ratio of the number of informative propositions gained per total task minute

(IP/TTM). Hoffman evaluated a variety of methods in studies of civil engineers who were experts at aerial photo interpretation. What his study showed was that unstructured interviews are indeed highly inefficient, having an IP/TTM value on the order of 0.15. Owing to the labor-intensive nature of protocol analysis, the TAPS method (see chapter 6) was similarly inefficient for the purposes of knowledge elicitation. A structured interview yielded an IP/TTM ratio of 1.0. A think-aloud task using "tough cases" was more efficient still, having an IP/TTM ratio between 1.0 and 2.0. This accords with the notion that the method of tough cases might reveal aspects of knowledge not readily found in documentation.

At the University of Nottingham, Nigel Shadbolt and his colleagues (Schweickert et al. 1987; Shadbolt and Burton 1990a) conducted a similar series of experiments at about the same time as Hoffman. They used some of the same methods that Hoffman used, although their constrained processing task involved concept identification and clustering task (card sorting). Nevertheless, their findings generally dovetailed with Hoffman's results in terms of method efficiency.

The two sets of studies also dovetailed in the sense that all of the methods seemed to allow the experts to express their knowledge about domain concepts and about procedures. This spoke to the issue of "differential access," that different methods for knowledge elicitation might differentially access forms of knowledge that are somehow distinct (Hoffman et al. 1995).

A subsequent study in the domain of weather forecasting (Hoffman, Coffey, and Ford 2000) involved the comparative analysis of the Critical Decision Method (CDM), the Recent Case Walkthrough, workplace and work patterns observations, and a number of other CTA methods. It was shown that Concept Mapping has a relatively high yield, with an IP/TTM of about 2.0. In addition, Hoffman also calculated efficiency ratios for the yield of leverage points, which were defined as aspects of the work domain or work context (according to the senior experts) where an infusion of new or better technology might lead to a performance improvement. Hoffman found that the CDM and workplace observations were relatively the more efficient methods for the identification of leverage points. At the same time, all of the various methods identified leverage points in all of the categories Hoffman et al. used (i.e., work procedures, knowledge sharing, workspace design, decision-aiding, and so on).

The results provide a useful qualification to previous reports on the CDM (e.g., Hoffman, Coffey, and Ford 2000). For the study of weather forecasters, each CDM session had to span more than one day. On the first day the researcher would conduct the first three steps in the CDM, then retreat to the lab to input the results into the method boilerplate forms. The researcher returned to the workplace on a subsequent day to complete the procedure. Weather forecasting cases are rich (since weather phenomena span days and involve dozens of data types and data fields), and forecasters' memories of cases are often remarkably rich. Indeed, there is a tradition in meteorology to convey important lessons by means of case reports (e.g., Buckley and Leslie 2000). Hence, the CDM worked quite well in this research project as a method for generating rich case studies. The impact of this domain feature was that the conduct of the CDM was time-consuming and effortful because the researchers sought exhaustive documentation for the incidents. Past measurements had suggested that CDM sessions take about two hours. This study involved a more inclusive measure of effort—total task time—and the CDM took about ten hours per case.

We see in these studies how measures can be created and tailored to address particular issues in CTA methodology and to explore those issues in the cognitive field research setting. In addition, this work shows that measures can be used to guide the choice of CTA methods and to guide the process of using CTA to inform the design of decision aids (for a review, see Hoffman and Lintern 2006).

Summary

Measurement of various kinds and combinations is critical in shaping CTA, conducting CTA, and going from CTA results to conclusions and recommendations. Cognitive Task Analysis methods are well suited to identify cognitive requirements and to support the creation and application of cognitive measures to study macrocognitive functions. With the growing use of information technologies, developers need to determine how their systems are affecting cognitive processes. Cognitive measures are relevant to software support systems. Cognitive measures are also needed with the variety of training programs aimed at improving decision making and other macrocognitive functions. We invite all of these communities to borrow and apply CTA probes and methods as they find practical. Practitioners may want to conduct CTA studies, or they may just need to tap into cognitive requirements in a piecemeal fashion. Whichever is the case, deploying CTA methods in the evaluation process should help produce results that lead to better system performance.

15 Future Directions for Cognitive Task Analysis

We can consider knowledge, expertise, and cognitive skills as valuable *resources* that we can tap into via Cognitive Task Analysis (CTA). We can treat CTA studies as explorations to determine how to support people. The knowledge gained from a CTA study is a product we can use to support specific target groups. In this way we can chart out a CTA research agenda using the concept of resource management.

Let's draw an analogy to another discipline that is concerned with resource management. Consider petroleum as a valuable resource. The petroleum engineering industry has developed techniques for locating it, assaying it, extracting it, processing it, applying it, and preparing workers to engage in these activities. Journals are devoted to best practices, conferences are held on innovative ideas, and training programs are conducted to upgrade skill levels.

The identification, elicitation, and application of expertise can be treated similarly. In order to make progress, we need to expand the cognitive models we have, just as petroleum geologists seek to learn more about the location of oil reserves. We need to expand the CTA methods we have, just as drilling experts seek to invent new techniques to extract and refine petroleum. We need to expand the applications of CTA, just as chemists and engineers search for new ways to use hydrocarbons beyond fuel. And we need to support expanded competence of practitioners, just as the petrochemical industry seeks to disseminate information and provide specialized training.

By treating people's strategies and knowledge and models as resources, we can seek better ways to capture what people know and believe, better ways to apply this knowledge, and better ways to increase the skill level in our field.

The resource management metaphor may help researchers take a more appreciative stance towards the participants they study. Instead of seeing them as "subjects" who generate data, we can regard them as collaborators in a joint venture of discovery. Many of them carry specialized knowledge that we can use. The experts have learned tricks of the trade that are rarely described in procedures manuals. The novices are relying on strategies and mental models they have learned elsewhere. They are engaged in their own informal quest to make sense of a task, even though they aren't scientists. If people continually make mistakes, we can be amused at their foibles, or we can wonder what is actually going on—perhaps there is something we're missing.

One researcher, well trained in laboratory methods, tried to use CTA techniques but was unable to ask follow-up questions. He needed to ask every person in his study the same questions in the same order, regardless of how they answered his questions. Another researcher, in a different project, noticed that many of her participants were showing emotional reactions to the survey questions she was using. But she couldn't bring herself to ask them about it. When colleagues suggested that she ask them after collecting their data, she replied that she hadn't been including this question from the start, so the findings wouldn't be scientific. She couldn't bring herself to engage the participants in conversation.

We hope researchers will be able to treat the people in their studies as participants, rather than as objects—or, in this case, subjects. We can learn a great deal from engaging in a respectful inquiry. The CTA methods we have described in this book can serve as a vehicle for such a dialog.

By treating research participants as resources, we can pursue our research agenda to expand our models, our methods, our applications for CTA, and our competence in performing CTA studies.

Expanding the Models

In chapter 8 we presented the framework of macrocognition and identified specific models that have been presented for each of the macrocognitive functions. Model development is a useful starting point. It helps us as researchers to keep our attention focused on the phenomena—on the cognitive functions and processes we are trying to understand. But we need to go further.

Additional Functions

We need to add to the functions and processes discussed in chapter 8. Surely there are other macrocognitive functions that affect the way people think when performing natural types of work; or perhaps some of the functions and processes can be combined. We can support cognitive work only if we understand what types of functions people are carrying out.

Stronger Models

We need to improve the models we already have. Some of the models are reasonably specific and can be used to make testable predictions and also to help us carry out applications such as designing an interface. But some of the models are still fairly shallow. They are little more than lists represented in diagrammatic form, capturing the ideas that are already understood without hypothesizing about causal relationships and without making any nonobvious claims. One of the values of science is testability, which means that theories have to be falsifiable. Our theories also have to have information value—they have to take a stand about which of the alternate accounts of a phenomenon to adopt. If our models avoid making any assertions that can be wrong, then they are not contributing to scientific progress.

Researchers conducting CTA studies need clearer and more comprehensive descriptions—models and theories—of the types of cognition they are going to encounter. Researchers also need to appreciate the limitations of existing theories so they can use CTA to probe more deeply into the phenomena they study, thereby improving the existing models. Two examples of this are the need for richer accounts of mental models, and the need for better descriptions of goal-directed thinking.

How are mental models formed, and how do they work? The concept of mental models is frequently invoked in CTA reports, but researchers rarely examine what mental models are. For the most part, the CTA community is content to rely on the experimental treatments of mental models (e.g., Gentner and Stevens 1983; Johnson-Laird 1983). These treatments were based on experiments using laboratory tasks. Can we synthesize field research to produce a naturalistic account of mental models? For some field researchers, the very idea of studying mental models may seem too amorphous.

Hoffman (1991) has developed an innovative technique for examining models of reasoning. In a collaborative interview, Hoffman worked with each of a number of forecasters to create flow diagrams depicting how they reasoned during forecasting. Entries included such things as "form a hypothesis" and "compare to computer forecast." After some months delay, Hoffman returned to the forecasting officer and presented all of the model diagrams to the participants, along with some "bogus" models and models of apprentice reasoning. In that task, participants provided useful information about strategic reasoning, reasoning styles and ways in which the forecasters shared (or failed to share) their reasoning heuristics.

How do people engage in goal-directed thinking? Field researchers often invoke the concept of goal hierarchies. We all appreciate that the goals of decision makers have a very important influence on their behaviors. Field researchers have developed various methods for mapping goal hierarchies, ranging from the Abstraction Hierarchy (Rasmussen 1985; Vicente 2002) to Hierarchical Task Analysis (Annett 1996; Shepherd 2000) and its variant Goal-Directed Task Analyses (Endsley, Bolte, and Jones 2003). Useful as these methods are, they are means of representing goal-related data, rather than the reflection of how people think about goals. The mapping of goals into hierarchies gives the impression of orderliness—everything is in its place. However, Elm et al. (2003) have argued that hierarchies are too simplistic. Elm et al. described a formalism for using networks (not unlike concept maps) to reflect the richness of goal interactions. We would like to take this further—anytime a person or organization has more than one goal, which is just about all of the time, there is the potential for goal conflicts. CTA studies are needed to provide useful accounts of how people and organizations resolve goal conflicts, how they specify goals in order to prevent these conflicts, and how they make constructive use of the goal conflicts.

Teams and Organizations

We need to examine the way teams and organizations perform macrocognitive functions. Are there emergent strategies at the team or organizational level? Are there unique barriers not encountered by individuals? For the most part, our macrocognitive models have concentrated on the thinking of individuals.

Expert/Novice Differences

We need to develop better descriptions of expert/novice differences. Currently, we represent expert/novice differences as contrasting lists of attributes. We have to push further than this. In many field settings expert performance is the "gold standard." Decision researchers have traditionally looked for the boundaries of expertise and the ways in which people's decision making can be compromised. Expert system developers have treated expertise as a collection of production rules. The CTA community views expertise as a complex phenomenon. Our accounts of expertise need to reflect this complexity, describing how expertise develops, the range of conditions favoring rapid or slow development, and the boundary conditions for transfer of expertise. Weiss and Shanteau (2003) and their colleagues have developed a measure of expertise that reflects accuracy and consistency. We hope other researchers will adopt it or improve on it.

Expanding the Methods

Defining Boundary Conditions of Existing Methods

We need to compile our lessons learned as a field in order to determine the boundary conditions for various CTA methods. Currently, different research groups develop their own methods and are defensive if these are criticized. Unless we can foster a constructive dialog about the strengths and limitations of different methods we will have trouble making progress. Hoffman's (1991) CTA research on weather forecasters used converging operations from different CTA methods, and this is one way to obtain comparisons and determine what different methods have to offer. Ross et al. (2003) arranged for different research teams to observe the same Army exercise, and this is another way to achieve a comparison of methods.

Expanding Existing CTA Methods

Our current CTA methods are just a starting point for conducting cognitive field research. In the coming years we expect to see more and more researchers adapting existing methods as they apply CTA to different projects. William Wong's work is an exciting example of advances in CTA research. He and his colleagues have modified the CDM in a number of ways, including expanding the probes (O'Hare, Wiggins, Williams, and Wong 1998); using stick-on notes to facilitate dialog (O'Hare et al. 1998); developing an approach to thematic analysis of CTA data, the Emergent Themes Analysis (Wong 2004; Wong and Blandford 2004); and designing new forms of representation for CDM interviews such as decision charts and decision analysis tables (Wong 2004). These expansions make the CDM more structured; in particular, they make the data analysis process more directed. Wong, Sallis, and O'Hare (1997) describe ways of using CDM findings to create display design concepts. Klein and Armstrong (2004) have also described some ways that researchers have modified the CDM in practice.

Creating New CTA Methods

In order to conduct in-depth studies of the various macrocognitive functions and processes, we will need to adapt CTA methods that were originally developed for other purposes, and we will need to create new methods. Researchers often find it easy to persevere with familiar methods rather than being eclectic and adopting the method most suited to the question. Mastering new methods requires time and effort. However, we will be limited in the range of phenomena we can investigate unless we are prepared to innovate and to explore new methods for knowledge elicitation, analysis, and representation. Most CTA methods are fairly labor-intensive. Militello and Hutton (1994, 1998) described a set of applied CTA methods that did reduce some of the workload, but the field needs more progress here. The challenge is to find ways to retain richness and depth of the data, not just spend less time.

Computational Modeling

Computational modeling of macrocognitive functions is becoming more common. Chandrasekaran and Josephson (2000), Sokolowski (2003), and Gonzalez, Juarez, and

Graham (2004) have been modeling situation awareness. Warwick and colleagues (Warwick et al. 2002; Warwick et al. 2001; Warwick, McIlwaine, and Hutton 2002) have been developing computational versions of the RPD model. These models are primarily useful for improving the quality of computer-generated forces and are fairly weak in describing the actual decision making of experienced commanders. Nevertheless, the process of trying to capture these functions in a computational form is a valuable discipline and can be of great benefit in improving the quality of our conceptual models (e.g., Warwick and Hutton, in press).

Uses of Stories

Researchers may find it instructive to explore the use of stories as means of conducting CTA studies and representing the findings. In order to take advantage of the impact of stories, we need to define what counts as a story and how best to structure stories. Here the cognitive field research community will have much to learn from other disciplines that have been studying storytelling for many years (Brown et al. 2004; Denning 2005; Frank 2002; Mattingly 1991; Mostert, Zacharkiewicz, and Fossey 1996).

Challenges to CTA

We will need to confront direct challenges to the CTA methods we have been using. One challenge has been mounted by Vicente (1999), Roth, Patterson, and Mumaw (2002), and others who rely on the cognitive work analysis (CWA) perspective. They have taken the lead in trying to capture the constraints in the workplace, independent of the cognition. They assert that experts' mental models are often flawed and argue that new information technologies should not be designed on the basis of knowledge elicitation results. They have criticized CTA studies that examine only the thinking of the participants and fail to describe the context in which this thinking occurs. We disagree with their reluctance to take advantage of the insights that can be gained from subject-matter experts. Their claim that cognitive work can be supported by studying the workplace independent of the decision makers, seems as artificial as studying decision makers without considering how the workplace affects their cognition. But we do agree with them about the value of capturing workplace features. Their efforts should benefit the CTA community by providing us all with techniques for characterizing the work environment.

However, the CWA movement sometimes appears to study workplace dynamics in opposition to cognitive dynamics rather than in combination with cognition. The emphasis on setting follows Gibson's lead (1957), which has encouraged ecological psychologists to look for intelligence in the world and not in the mind. Some CWA proponents similarly have sought to avoid "mentalism" and explain phenomena entirely through descriptions of the constraints in the environment (also see Mace 1977). While we admire these efforts, they sometimes appear to be a retreat into the antimentalism views of the behaviorists. The CWA challenge for CTA researchers is to study contextual factors along with cognitive ones. If methods developed by CWA researchers are too cumbersome—for example, the abstraction-decomposition matrix—then we will need to construct simpler methods.

The concept of "distributed cognition" may be helpful in breaking researchers free of a fixation on cognition as happening only between the ears. Distributed cognition takes the view that we think with our minds and with our surroundings, using artifacts to manage attention and memory. The study of cognition, therefore, needs to be sensitive to the materials at hand. For example, the description of macrocognition in chapter 8 explicitly states that these functions might be performed by individuals, or by teams interacting with available technologies.

Another challenge to CTA is to sort out the kinds of distortions the methods may introduce. For example, Feltovich et al. (2004) speculated that the act of abstracting qualities of events, a reductive tendency, generally leads to oversimplification. Abstraction is, by definition, a simplification. Yet when we use CTA methods we are usually trying to abstract cues from the flow of experience. Should we worry about the distortions we are introducing? Wilson (2002) and Wilson and Schooler (1991) have demonstrated that by asking people to verbalize their experiences, we are increasing the risk of distortion. The act of verbalization has this effect. Should we worry about asking our SMEs to verbalize? Loftus (1996) has demonstrated how easy it is to create inaccurate memories in research participants. Should we worry about the demand characteristics of our methods? Are there safeguards we can introduce to minimize these types of distortions? Ericsson and Simon (1984) developed safeguards for protocol analysis. Should we adopt their safeguards, or develop similar ones? In chapter 6 we dismissed these concerns because our experience has been that skilled participants are not as malleable as the participants in some of these studies. We have not found reason to worry about systematic biases. Nevertheless, we have also learned not to believe everything that our research participants tell us. One of the reasons for using a timeline in CDM interviews is to catch inconsistencies in the way people present their incident accounts. All of these issues warrant further research to examine the accuracy of incident recall of skilled professionals, and to devise techniques to reduce errors.

This raises the question of how pristine our methods need to be. On the one hand, by adopting the stringent criteria used by Ericsson and Simon we avoid many of the

problems of distortion because their protocol analysis simply recorded what participants said and did not introduce questions or probes to get fuller explanations. On
the other hand, this self-imposed restriction seems inappropriate for naturalistic research. It makes sense for collecting data under carefully controlled conditions, but it
does not make sense for learning new things about phenomena of interest. Blumer
(1969) has warned us to be on guard against methodological imperialism—setting up
the method as more important than the phenomena we want to study.

Expanding the Applications

The five levers for improving performance are technology, training, organizational design, selection, and incentives. CTA can improve the use of each of these levers.

Technology

The Decision-Centered Design (DCD) approach was specifically developed to use CTA
as a basis for developing better systems, especially information technology systems.
There are additional ways to apply CTA for enhancing technology. The field of knowledge management has many of the same goals as DCD—to acquire critical aspects of
knowledge and make these more readily available to the organization. Cognitive Task
Analysis methods could strengthen the knowledge management field by providing
tools for learning more from SMEs, tools for analyzing and representing knowledge,
and tools for understanding how people in an organization will use the material in
knowledge management systems so that these systems can be better designed. One of
the bottlenecks of knowledge management is the work of extracting information from
experts. Here, auto-CTA techniques could be useful, if we can figure out how to help
people elicit useful information on their own. Another knowledge management bottleneck is storing and retrieving the information. Story or vignette formats may improve
access and use of the information. In addition, CTA representations help system developers to understand the way their systems are supposed to work in the field. These representations can help to bridge the gap between cognitive systems engineers and
software developers.

Training

Chapter 12 showed how CTA could be applied to training. Cognitive Task Analysis
methods can be widely incorporated into the field of education as general-purpose
tools for conducting inquiries and learning from SMEs. Much of education is structured as a process that is one step removed from the phenomena of interest—reading

articles about how tasks are performed and skills are applied, rather than observing and interviewing the practitioners of those skills. The burden of education is on the instructors to find effective ways to convey what they know, rather than on the trainees to elicit the subtle aspects of expertise that teachers may gloss over. Cognitive Task Analysis methods may be useful for the instructors themselves, to sensitize them to these subtle, perceptual skills. Too often, classroom preparations cover the procedures rather than the perceptual discriminations. In the early development of Applied Cognitive Task Analysis (ACTA), we worked with Navy electronic warfare coordinators (Crandall et al. 1994). One of the SMEs participated in a two-hour CTA interview in which he was asked a wide range of questions to probe the nonprocedural knowledge he possessed. He later reported that he found these questions challenging and even frustrating because he didn't have any pat answers. He was irritated when he walked out of the interview, but he noticed that the next time he taught his course at the school, he was explaining subtle aspects of performance that he had never tried to cover before. As a result, he became an advocate for using CTA to help instructors teach the difficult aspects of their jobs. This example illustrates how CTA methods can help trainers do a better job of imparting knowledge.

CTA methods can also be used outside of the classroom, particularly in on-the-job training (OJT) situations. Chi (1996) has argued against the usual practice in OJT and tutoring sessions of having the SME convey information to the trainee. Chi's point is that it is better instructional practice to have the trainee take an active role in the process, even directing the process by probing the SME. Pliske et al. (2000) developed a program for helping SMEs articulate the subtle aspects of their expertise in OJT settings and having the trainees learn to ask better questions in order to put Chi's advice into practice.

Finally, as organizations become larger they struggle to distribute the lessons that individual workers have learned. CTA can be valuable for knowledge management—to elicit stories and viewpoints, and to format these so that others can learn from them.

Organizational Design

Team CTA methods can be valuable for understanding how organizations can assign tasks to improve coordination and reduce coordination costs. Klinger and Klein (1999) described a project that used team CTA to redesign the emergency response rooms of nuclear power plants. Team CTA has also been applied to emergency rooms in hospitals, operations centers on board Navy ships, Marine Corps command posts, and other settings. The driving force for applying team CTA to organizational analysis and design is to understand the tradeoffs and compromises made in dividing and sequencing tasks. Organizations have to apportion responsibilities to teams, and every method of apportionment requires tradeoffs and incurs coordination costs. Furthermore, over time organizations accumulate inefficiencies, as obsolete tasks and roles are retained because no one realizes that they are no longer serving any useful function. By using team CTA to delve into these tradeoff and coordination costs, we can identify unnecessary barriers as well as strategies to improve the accomplishment of macrocognitive functions at the team/organizational level.

Personnel Selection

As more jobs take on cognitive and decision requirements, better methods are needed to assess the capability of applicants. Personnel selection batteries are easier to devise for physical and procedural jobs than for cognitive tasks. What is going to happen if we introduce macrocognitive functions into the field of personnel selection? Will we be able to formulate CTA methods that can be sufficiently sensitive to capture macrocognitive functions, while being sufficiently robust to withstand legal challenges?

Incentives

Managers struggle to find ways to motivate their subordinates. Several researchers have been skeptical of the impact of incentive systems. One barrier to incentives is that workers may fail to understand what they mean. In one CTA project (Klein, Wiggins, and Green 2000), we interviewed software developers at a major corporation that provided annual salary increases of 6 percent. However, after probing we found that some employees believed that this only matched the rate of inflation. In fact, at the time the rate of inflation was less than 1.5 percent. Employees concluded that their raises were merely keeping up with inflation, rather than quadrupling these rates. If managers don't understand the way incentives are perceived, they will be hard pressed to use incentives effectively.

Expanding the Competence Level

We want to develop "CTA masters," individuals who are recognized as embodying the best practices in the field. We also want to prepare more practitioners in the research and development community who can conduct effective CTA studies, even if they aren't CTA masters. If we achieve these goals, the quality of CTA studies will improve, and we will learn more from our projects. The pursuit of these goals will require better training, more meaningful standards, and broader dissemination of CTA research.

Training CTA Practitioners

We have a training challenge to improve the quality of CTA skills. The challenge is to train more people and to train people to higher levels of ability. It should not be hard to improve the existing training, because there is virtually no available training in CTA. The applied researcher who wants to acquire CTA skills will have difficulty figuring out how to accomplish that goal. Perhaps two to three brief CTA workshops are offered in a given year (a few days long at most) and these have a limited enrollment in order to provide more hands-on experience. In academic settings, courses are offered in experimental methods and textbooks are available, but few, if any, graduate programs offer courses in CTA. No CTA textbooks are available, and few faculty members are prepared to teach such courses. Some of the people who perform CTA studies in the field acquire their skills by pairing with a practitioner who is experienced in performing CTA or by reading selected articles and chapters.

On the positive side, Militello, Hutton, and Miller (1996) have produced a CD-ROM that provides familiarization with a few CTA methods. And if we broaden the search we do find a number of classes in qualitative methods in a variety of disciplines. Therefore, we need to both develop more instructional materials and synthesize the materials that have been developed.

Perhaps we need CTA studies comparing the way experts and novices perform CTA. One reason we have difficulty in knowing how to train the subtle skills needed to conduct a CTA study is that we have not contrasted experienced CTA researchers with novices. We haven't tried to pinpoint the types of knowledge and techniques that enable seasoned CTA workers to ask the right question in the right way, or pick up on the hesitation or the invitation to probe further.

Should we consider individual differences in ability to perform CTA? Some trainees seem to pick up CTA skills quickly, even the most demanding of the techniques. Others never seem to get comfortable with doing qualitative, naturalistic research. In using incident-based CTA methods, we have suspected that some people may not have a good sense of what constitutes a story. They have trouble hearing the stories that SMEs are telling, and they have trouble telling these stories to document their findings. Are there individual differences related to personality and prior experience that affect whether someone will be outstanding at doing CTA, or will struggle with the methods and come to avoid them? If so, can we pre-test for these and screen out people who are not prepared to enter into the thinking of someone else?

Can we design simulations for CTA training? These simulations might use case studies to describe a verbal record, in order to see where trainees would follow up with questions, what types of things they would ask, how they would continue when faced with certain answers, and how they would then diagram the material so that others could understand it. The trainees might be able to tap into the actual follow-up questions posed by experienced CTA researchers, perhaps including the rationale for these questions.

Standards

We have a challenge to define CTA standards. What counts as a good CTA study or a poor one (Hoffman and Woods 2000)? Klein and Militello (2004) suggested a few criteria: To be considered successful, the CTA study should result in an important discovery about key judgments and decisions and macrocognitive functions, it should effectively communicate that discovery, and it should have an impact on the sponsors. These criteria are about outcomes. We need similar criteria about processes, about the way CTA studies are performed. Researchers will need to engage in debates about procedures used in prior studies, reviewing the methods and suggesting better ones, and arguing about their feasibility and benefits. Even if we are not comfortable criticizing the work of others, we can at least identify CTA projects that can serve as exemplars of best practices. That way, we can use connotative as well as denotative definitions of success—defining success through example as well as through explicit criteria.

Dissemination

No scientific field can flourish without the means of disseminating and debating its findings. By facilitating the process of dissemination we will facilitate the growth of competence in performing cognitive field studies. Traditional journals are becoming more open to publishing CTA studies, and specialized journals catering to CTA studies are also appearing. Journals are critical to CTA development because they require research standards for editorial reviews. But print journals are not an entirely satisfactory vehicle because cognitive field research usually leaves a trail of transcripts and observation notes that cannot be packaged into limited journal space. Electronic archives seem to be needed, but these pose difficulties of their own about hosting and managing the archives. If we can find ways to overcome these barriers, we may see more secondary research projects—projects that reuse documented materials from earlier studies in order to avoid the time and expense of collecting new data.

We need newsletters and clearinghouses that permit researchers to quickly check to see who has already done work with a method, or in a domain, or on a macrocognitive phenomenon. Information exchange depends on informal networks: "Oh, you are studying XXX—you ought to talk to the folks at YYY laboratory, they were doing some interviews in that area a few years ago." Informal networks are important, but we might also benefit from more comprehensive mechanisms. And databases are not the answer either. People may not search databases unless they have reasonable confidence of finding something useful. A medium such as a newsletter would allow researchers to announce the completion of reports or articles by posting a few paragraphs, an abstract, or an executive summary.

Finally, conferences offer an essential means of disseminating CTA research. However, giving presentations at conferences is necessary but not sufficient. Meaningful conference interchanges include appraisal of methods used, probing of findings reported to gauge their robustness and their implications, and attempts to use data to falsify models. Conference presentations need to center around the data that were collected because data and findings are the basis of scientific advancement. One of the dangers of naturalistic research is that researchers can find it easy to submerge data underneath a flood of speculations and opinions. Scientific progress is marked by the accumulation of lawful relationships, particularly nonobvious relationships, rather than by the unveiling of new models. In accepting papers for conferences we should emphasize the lawful relationships that have been identified by CTA studies.

Taken together, these recommendations suggest how we can strengthen our CTA competence through improved dissemination. The steps we propose—increased journal publication, accessibility to the documentation following project completion, development and use of clearinghouses and newsletters, and forums for debate—can help us inform and transform cognitive field research and the uses and applications of CTA.

Summary

We hope that throughout the reading of this book you have learned some useful methods along with some effective ways of implementing these methods. We have tried to describe our trade secrets, to help you understand various CTA methods and to give you the courage to explore new approaches to gathering data. We have tried to show the depth, richness, and value of CTA and its applications. As you can see, there are no hidden mysteries, only hard work and practice, as with mastering all complex skills.

Appendix: Guidance for Data Collection

This appendix has a number of purposes. One is to describe some useful Cognitive Task Analysis (CTA) methods that were mentioned but not detailed in the preceding chapters. Another is to show, by providing sample research protocols and forms, what exactly happens in conducting CTA procedures. The goal is to demystify CTA and encourage individuals who might want to conduct CTA procedures. Our presentation includes protocol notes. These are ideas, lessons learned, and cautionary tales for the researcher. The guidance offered here spans the gamut from general advice about record-keeping to detailed advice about audio recording knowledge elicitation interviews.

To provide additional examples and guidance, we present some templates—prepared data collection forms for selected procedures. Digital versions of the methods protocols and templates are available for download as either MS Word documents or as PDF files from www.ihmc.us. In the "Establishing Rapport" section that follows, we provide guidance and lessons learned on how to create that situation. We also provide guidance about pre–data collection activities, conducting interviews and observations, and recording data.

This appendix describes generic guidelines for interviews and observations and how to combine them. Some of the tips imply data collection events that were carefully planned; but just as important as scheduled data collection are unexpected opportunities to collect data. For example, during the weather forecasting case study (Hoffman, Coffey, and Ford 2000), a group of pilots was effectively stranded at the airfield owing to inclement weather. The pilots were gathered chatting in a hallway. This was taken as an opportunity to conduct an informal and unplanned interview about weather impact on air operations. At any moment there may be an opportunity to ask a few probe questions, take notes about work patterns (e.g., information sharing), or make an observation that suggests leverage points.

Documentation Analysis

The process of coming up to speed typically involves an analysis of documents (manuals, texts, technical reports, etc.). The literature of research and applications of knowledge elicitation describes documentation analysis as an important and necessary method (Hoffman 1987; Hoffman et al. 1995). Documentation analysis is relied upon heavily in the work domain analysis phase of Cognitive Work Analysis (see Hoffman and Lintern 2006; Vicente 1999).

Documentation analysis can be a time-consuming process, but it can also be indispensable in knowledge elicitation (Kolodner 1993). For example, in a study of aerial photo interpreters (Hoffman 1987), interviews about the process of terrain analysis began only after an analysis of the readily available basic knowledge of concepts and definitions. To take up the expert's time by asking questions such as "What is limestone?" would have made no sense.

Although it is usually considered part of preparation, documentation analysis invariably occurs throughout the entire research program. For example, in the weather forecasting case study (Hoffman, Coffey, and Ford 2000), the preparation process focused on analyzing published literatures and technical reports that referenced the cognition of forecasters. That process resulted in the creation of the project guidance document. However, documentation analyses of other types of information (records of weather forecasting case studies, standard operating procedures documents, local forecasting handbooks, and so on) occurred throughout the remainder of the project.

Analysis of documents may range from underlining and notetaking to detailed propositional or content analysis according to functional categories. This type of documentation analysis may involve specific procedures that generate records of the knowledge contained in the documents. A tip here is to record these references as you read them, which will avoid a frenzied search for source details when you are trying to finish a report. Documentation analysis can be useful for a variety of purposes, including:

- Contributing to the development of models of reasoning for the domain of interest.
- Construction of knowledge models, since the literature may include both useful categories for, and specific examples of, domain knowledge.
- Identification of leverage points—aspects of the work where even a modest improvement in technology might result in a proportionately greater improvement in the work.

Documentation analysis also has its limitations. Practitioners possess knowledge and strategies that do not appear in the documents and task descriptions and, in some cases, could not possibly be documented. Generally speaking, people who work in complex sociotechnical contexts possess knowledge and reasoning strategies that are not captured in existing procedures and standard documents (McDonald, Corrigan, and Ward 2002). The only way to learn about that knowledge and those strategies is to collect data.

The Interview

CTA interviews might be conducted at the participant's workplace, at the researcher's workplace, or at some other location. We have conducted interviews in airplane hangers, the backseat of cars, in fast food restaurants, fire stations, aboard ships, in nurses' stations—you have to be prepared to adapt wherever you end up, and make the interview work.

There are two aspects to any interview: (1) getting the information that you are there to obtain, and (2) handling the procedural and interpersonal aspects of the interview. Here we offer guidance and lessons learned regarding the procedural and interpersonal components of skilled interviewing.

In laboratory experiments, it is important for the researcher to be a neutral presence, and to present the same persona to every participant—same instructions, same tone of voice, same pleasant facial expression. The last thing you want is to become a contaminating variable in the experimental design! Part of the training for doing experimental work is to learn how to maintain the same demeanor over the course of data collection.

In the type of CTA we are describing, the researcher has a quite different challenge. You are going to ask people to tell you about their personal experiences, in some cases, challenging experiences where things have gone smoothly. These may be events where lives were at risk, either the participant's own or people that she or he was responsible for. Asking a person to open up to you, a stranger, in this way requires a significant degree of trust. You must find a way to connect with the person you are there to interview and create that sense of trust. Good communication skills, active listening, and body language that conveys that you are eager to learn about them and their work—these elements will not ensure great data, but they are essential to creating a situation in which you have the chance to get great data.

Cognitive Task Analysis interviews are structured in that the researchers have a good idea of the information they want to elicit. They have prepared the probe questions and forms, and have in mind a sequence for moving through the interview. But the structuring does not require that questions be posed exactly the same way each time or in a single order. The well-conducted CTA interview emerges from the interaction between the participant and the interviewers. For example, if topic #3 on the interview happens to come up in the middle of discussion of topic #1, let it. Don't waste time asking about it again after you are finished with topic #2. The researcher wants to engage the participant in a conversation rather than a rapid-fire sequence of yes/no, agree/disagree, survey-like items.

Skilled CTA interviewing requires patience, experience, and practice. Some people seem to do it well from the beginning; others require concentrated effort to hone the necessary skills. Each interview is a chance to gain feedback from your interview partner, benefit from lessons learned, and note effective questions and strategies for next time. We have identified a set of skills that the best CTA practitioners routinely do well:

- **Establish rapport** with the participant
- **Know how to use time to greatest impact**—when to spend time on a topic and when to move on
- **Ask good opening questions**: they know what they want and how to get it efficiently
- **Recognize where to drill down**, what to deepen on; they can hear words or phrases that flag rich areas for probing
- **Recognize when a direction is not fruitful**: they can turn the interview around and get it going in a productive direction
- **Recognize how to frame questions**: how to pose a question so that it makes sense but doesn't lead the interviewee
- **Recognize where the important content is likely to be** and how to bring the interview to that place (this may require a whole series of questions that set up a key insight)
- **Know how to reorient the interview** when the current direction isn't working
- **Have a range of strategies for probing**, deepening into content
- **Understand the power of silence**: they can wait for the interviewee to think about a question without filling the void with talk

In the sections that follow, we offer some of what we've learned about how to develop these skills and create a positive, productive, successful interview. Many of these suggestions are relevant for interviewing in general; others are more specific to incident-based CTA methods such as CDM. The guidance here assumes a two-person team interviewing a single individual, but it certainly can be applied to other interview situations.

Teaming

We typically interview in pairs for a number of reasons. Good data-collection sessions are dense with information. Managing the interview, keeping track of data quality, and capturing the interview in notes is a lot for one person handle. The quality of data is simply better when two people are present, particularly if they have defined their roles ahead of time as the Lead and the Second.

The Lead

Ahead of the interview session, interview partners should decide together who is leading the interview. The Lead is primarily responsible for the procedural and interpersonal aspects of the interview. The Lead will make a lot of eye contact, know what was just said, establish the line of questioning, and evaluate when it's time to switch topics and when it's important to stick with an issue and get more information. The Lead sets and maintains the pace of the interview, and his or her notetaking must be secondary to managing the interview.

The Second

The other interviewer is the Second. The Second's task is to pay close attention to the data-gathering requirements of the interview. His or her task is to take excellent notes of the interview, record the timeline, and keep track of what data elements are still lacking at that point in the interview. Because the Second interacts somewhat less with the participant, he or she should be able to filter out the surface, conversational aspects of the interview and focus instead on creating a solid data record.

Team Dynamics

It is helpful to discuss handoffs ahead of time. Should the Second jump in with questions or hold them until there is a pause in the discourse? Jumping in tends to work best when the interviewers had experience working together as data collectors and each has a sense of the other's style. If the Lead prefers not to be interrupted, then he or she has the responsibility of finding frequent break points and inviting Second to ask questions and clarify meaning.

Sometimes the Lead role is determined by the interviewee. There are all kinds of interpersonal dynamics—gender and racial differences, status differences, personality styles—that can come into play. For his or her own reasons, the interviewee may choose to interact primarily (or exclusively) with one member of the team. You have to be ready to switch roles if it works out that way.

Establishing Rapport

When you encounter a new set of participants—within a new domain, new organization, or new segment within an organization—the first few interviews are absolutely crucial. Your first interviewees will tell their co-workers about the experience, and those reports will filter through the organization. Time spent making a good connection with the first several participants will have a positive payoff for the rest of your interviews. It will be easier to recruit participants, and they will be more at ease and ready to talk openly from the start of the session.

Similarly, time spent helping participants feel comfortable at the outset means that they are more likely to invest their time and energy in the session and more willing to come back or to make themselves available by phone or email for follow-up questions. They are more likely to refer others to you. Clearly, it is an investment that pays off.

Most people thoroughly enjoy these interviews. For many people, it is an unusual opportunity to talk about their work, what they know, and what they do well. Nonetheless, you have to be ready to deal with a variety of initial reactions from people. These can range from the participant communicating, "You are going to have to win me over before I am going to tell you anything—who the heck are you anyway?" to "Hi, I know about you and why you are here, and I'm ready to cooperate, so let's get on with it."

Some tips for establishing rapport:

- **Social pleasantries.** Tell them a little bit about yourself. Give them a chance to talk a little bit about themselves. How long have they been doing this job? Been busy at work lately? These social pleasantries begin to establish the trust required to do the task together. Humor helps immensely. Your body language will communicate your interest and openness, so be aware of it.

- **Use your credentials wisely.** It's best to be cautious about using professional credentials as a way of establishing your credibility. In some settings and work domains, academic degrees are admired and respected. In others, they are beside the point and can make you appear pretentious, arouse evaluation anxiety, place even more of a barrier between interviewer and participant, and backfire as a technique for establishing rapport.

- **Reassure them.** Emphasize that you are not there to evaluate their performance, or whether they are a good or poor decision maker, planner, problem solver, or whatever it is you are there to explore. Tell them there can be many ways to get to a correct outcome and that sometimes poor outcomes occur despite good decision making. Because participants often feel some degree of evaluation apprehension to begin with, you may find that you have to repeat this throughout the interview, to reassure the participant that you are not in any position to evaluate whether what they did in a particular incident was good or bad. Convince them you are there to learn from them.

- **Inform them.** Make sure they know the purpose of the interview, who is sponsoring it, and a little about your organization. This isn't the time for a marketing pitch or your biosketch, but a couple of sentences about your work and what the project is about provides necessary context. If the interviews have been arranged by others and you have not previously talked with the participants directly, do not count on them knowing why they have been asked (or told) to talk with you. Be prepared to provide that information. Give them a business card and invite them to call you after the interview if they have questions or want to talk anything over. (We learned this lesson the hard way when a particularly antagonistic firefighter finally revealed that he believed he was being fired, and that we were there to give him a termination notice.)

- **Be engaged.** There should be lots of eye contact, nodding, smiling, and saying, "How interesting!" Convince both yourself and the participant that this is the most interesting story you've heard in a long while. Even if you are the Second, it is still important to look up and make contact with the person occasionally. Avoid spending the interview with your head down, engrossed in your notes.

- **Affirm their expertise.** This might sound something like, "I am a cognitive psychologist (or, I work with psychologists) and I/we study expertise. I know very little about what you do—you are the expert here, and I need your help in order to understand what you know." The phrase "I need help getting inside the expert's head" is one way to communicate this idea. People often laugh at this and make jokes about psychologists. We all have a good laugh and everyone relaxes a bit. This is not only about establishing rapport with the SME. It is also about the appropriate stance as an interviewer. You are not there to trade stories or to show off your expertise about the field. This isn't the time to demonstrate how much you know or to make suggestions about how the SME might have handled the situation.

- **Let them talk.** Schedule interviews so that you can accommodate one story of theirs, even if the story is not so relevant to your purpose. You don't want to have to cut them off during the warm-up because your interviewing schedule is too tight. If at all possible, leave thirty minutes between interviews, fifteen minutes minimum. Realize that these "side" stories can give valuable insights into the culture and the work environment in which this person functions.

- **Don't let them take control.** The emphasis on their expertise gives participants a considerable degree of control in the interview, and it occasionally backfires. Deferring to their experience can be taken as permission to run away with the interview or to pontificate. They may decide to talk about a different topic from the one you are there to find out about, and you may find yourself struggling to redirect the conversational thread. Maintaining rapport means not cutting people off quickly or rudely. You may find it useful to say, "That's so interesting, I hope we have time to talk more about it later; right now I'd like to hear more about...." It can also be very helpful to ask, "And did that happen in this incident, the one you were describing for us?" to reorient the interview to the point of the interview.

Pacing and Flow

It is part of the Lead's task to monitor and guide the flow of talk and keep the focus on agreed-upon study goals. In a domain where the interview content is highly technical, the discussion very concentrated, or the content emotionally charged, it may be imperative to take breaks. When interviews are being conducted onsite, the pace can be wild, full of constant and unavoidable interruptions. In either case it is up to the interview team to maintain continuity, to keep track of the discussion, and get the interview back on point. Some tips for maintaining good pacing and flow:

- **Start with good questions.** Once the interview begins, the first set of questions you ask conveys information about the type of data you are interested in. Taking time at the outset to repeat or restate a question that hasn't been understood will allow you to get desired information more quickly and efficiently later on in the interview or in later sessions. Interviewees will know what you want and anticipate your questions. They will begin to volunteer information of the sort you are seeking.
- **Don't turn it into a conversation.** Once the interview is underway, you should be listening far more than you talk. We want to help people feel comfortable, and some degree of informal verbal give-and-take is part of maintaining rapport. However, we are there to hear their stories, not trade back and forth. These are not conversations even though they have a conversational tone.
- **Be ready to back off.** If the interview session becomes uncomfortably intense or emotional, you might take the interview off on a bit of digression, to joke around or just chat a bit. This provides a brief, informal "rest stop."
- **Redirect when necessary.** As mentioned above, participants can get off track (of what we're interested in—it's not off their track). You must be willing to listen and be interested in what feels like a sidetrack, but not let it go too far afield. One way to redirect is to take the interviewee back to an earlier point in the interview ("Let's go back for a minute to the point in the incident when ...") and then redirect the interview. Another way is to say, "I'm really interested in hearing about this, but I don't want to lose track of what we were talking about a few minutes ago. Could we go back to that for just a bit?"

▪ **Adjust the pace of the interview.** Sometimes it seems that an interview is going too fast or too slow. Ask yourself "too fast/slow for whom?" It may be that the participant's verbal style is simply a lot faster or slower than what you are used to or prefer. If the participant is a slow talker, there isn't much you can do. At least realize that you will eventually get the information you need. However, the problem may be that the participant is taking a lot of time explaining things you already know. Instead of cutting him or her off by saying, "I already know that," you can speed things up by letting some of your own knowledge slip out in comments, acronyms, jokes, and so on. Participants will often ask calibrating questions to figure out what you already know; they will often readjust their language and level of explanation in accord with what they think you can follow.

▪ **Slow the interviewee down if necessary.** You face a more serious problem if the participant's verbal style is very quick. If you aren't able to keep pace, you can ask them to slow down a bit (they can see you are trying to take notes). But constantly interrupting to ask them to slow down can disrupt their flow of thought so badly you risk losing good information. If the participant continues to talk faster than you can handle, the Lead should check to see if the Second is keeping up (e.g., he or she is nodding while taking notes and doesn't appear to be scrambling). If the Second appears to be handling the pace, then perhaps the Lead can concentrate on following the verbal flow and worry less about catching details in his or her notes.

Information Quality

The "heart" of CDM is the use of a series of questions about a specific incident or event that focuses participants' attention on certain aspects of their own cognition. The questions allow them to tell us about what they were doing and how they were doing it.

Every project has a core set of issues and questions or probes designed to get at that information. It is essential that data collectors have a thorough understanding of those issues before data collection begins. For semistructured formats such as CDM, interviewers do not have preset questions with specific wording that participants are required to answer. It becomes the interviewer's task to figure out how to get good responses to the particular topic of interest. To do that, you have to be ready to ask the same question in a variety of ways, because a probe that works well for one person may draw a blank from the next. When a probe doesn't elicit the information you need, you have to know the reason for asking the question in the first place, in order to come at the issue from another direction.

In CDM interviews, it is very common for participants to drift away from the specific case and shift into a generic discussion of how things usually are done or a tutorial about domain knowledge. One of the indicators that this has happened is a shift from first to third person pronouns (e.g., "You can always tell when things go wrong... "). One of your tasks is to keep the participant focused on the facts of the specific case. This does not mean we aren't interested in the interviewee's general knowledge—if it influenced this case, it is meaningful. But if what they are doing is giving you a tutorial on basic procedures, you should refocus the discussion back to the specific incident. You can easily do this by asking, "Is that what happened in this particular case? Is that what you did this time?" It is your task to steer the interview, but it's important that you do this with finesse and respect.

CTA interviews are explorations. When you pose questions you open a door and ask the interviewee to walk through it and tell you what's on the other side. If you ask leading questions—if the interview is spent confirming your views and opinions—you have not allowed this person to tell you what he or she knows. Try to avoid questions that elicit simple yes or no answers. You should also be on guard against questions that are really attempts to confirm your view of the situation. However, there are times when you want to summarize what you have heard or your understanding of the incident. When this is the case, it is important to give the participant permission to tell you that you are wrong. Phrases such as, "I don't want to put words in your mouth. Tell me if I've got this, and where I'm off base," invite the interviewee to offer correction.

Many participants seem to recall an event more clearly if they are able to sketch aspects of the event as they describe it. This is particularly so when they are describing dynamic or spatial elements. It has become part of our procedure put a pencil and paper out on the table. We may mention, "There is paper here in case you find it useful while we're talking," or we may say nothing at all. We do not give instructions or request that they draw anything. Often, in the midst of describing something of what happened, they will grab the paper and begin drawing the scene. This technique helps focus their attention and frees them from the normal social conventions that surround conversation. You are more likely to see cognitive strategy coming out when they are talking and sketching out some part of the incident for you.

One reason that it can be hard to get the information we want is that we are so often working at the edge of what people are able to articulate. A large part of the interviewer's task is to help people talk about what they know and how they know it. One often has to circle and circle around something, going at it in different ways to help people articulate it. Some of the strategies we use:

- **Rephrase your initial question.** People understand things differently. Simply asking the question in another way can sometimes do the trick.
- **Go back several steps in the sequence** they have just described to you and say it back to them. You can introduce this by saying something like, "Let me see if I understand what was going on at this point." Hearing it back seems to help them recall details of the next part of the sequence.
- **Ask them to visualize and relive** that moment "as though it's on video." Again, take them back a few steps, lead them up to the point in the incident that is stumping them. Pose your retelling in the present tense, as though it's happening in present time. Remind them of any visual or auditory cues they have mentioned. Bring them up to the point in the incident that is giving them trouble and say, "Okay, there you are—now, what do you see? Hear? What's happening right now?"
- **Watch the participant.** The spontaneity and assertiveness of the participants' response will tell you whether your probes are hitting the mark. Other cues are tone of voice and eye contact. The interviewee will talk with no wandering or distractions as thoughts rush in and their words rush out. If you are taking notes, you have to scramble to keep up because it comes out so quickly.
- **Watch yourself.** One of your tasks is to approach each participant and each interview fresh. Walk into each interview with the idea that there is something valuable to learn from this person. It may be the last interview in a series, but that person may say something that is a complete surprise, that you haven't heard from anyone else. Be ready to hear that new information.
- If it seems as though the interviewee has said everything there is to say, but you still have a nagging question or something just doesn't fit together, trust that sense and **keep asking**. Similarly, it is important that you allow participants to give you information that may be redundant, rather than cutting them off. Redundancy across study participants is not a negative or a waste of time. In fact, it may be a key finding that a certain type of data replicates across interviews.

Figure A.1 presents a template for conducting a CDM interview.

Workplace Observations and Interviews

Workplace analysis can focus on the workspaces in which individuals or groups work, the activities in which workers engage, the roles or jobs, the requirements for decisions, the requirements for activities, or the "standard operating procedures." These alternative ways of looking at workplaces will overlap. For example, in a particular work group,

CRITICAL DECISION METHOD

Interviewer(s)   Date   Participant

Incident Identification

Instructions:
- ⊙ Obtain an incident
- ⊙ Ask for an overview

Ask:
- Can you think of a time when you and your skills (type of incident) were really challenged?
- Tell me about the last time you...
- Can you think of a time when your skills really made a difference—maybe things would have gone differently if you weren't there?

Listen for: An incident that fits your study goals, in which your participant played a key role.

Timeline and Decision Point identification

Instructions:
- ⊙ Repeat back the incident
- ⊙ Construct a timeline or diagram
- ⊙ Record decision points, shifts in understanding, and major events
- ⊙ Ask clarifying questions

Ask:
- Do I have this right?
- Where on the timeline should I put this?

Listen for: Decision points, shifts in understanding, places to probe, gaps in the story, gaps in the timeline, conceptual leaps, anomalies/surprises, errors, ambiguous cues

Flags: I just knew... It felt right... I guess... It was just a gut feeling... Something felt wrong... I've seen it before... It depends...

Deepening

Instructions:
- ⊙ Ask questions until you understand the incident
- ⊙ Use the timeline for clarification
- ⊙ Repeat back confusing points

Ask:
- What was it about the situation that let you know what was going to happen?
- What was it about the situation that let you know what to do?
- What led up to this decision?
- What were your overriding concerns at that point?
- How would you summarize the situation at this point?
- What were you noticing at that point?
- What were you seeing, hearing, smelling?
- What information did you use in making this decision?
- How did you get this information?
- What knowledge was necessary or helpful in this situation or at this point?
- What were your specific goals at this time?
- What were you hoping/intending to accomplish at this point?

Listen for: Critical decisions, cues and their implications, ambiguous cues, strategies, anomalies/violated expectancies.

"What If" Queries

Instructions:
- ⊙ Use "what if" questions to tease out specific elements
- ⊙ Ask what a new person might have done
- ⊙ Ask what mistakes might have been made earlier in interviewee's career

Ask:
- Did you consider other alternatives?
- Might someone else in the same position have done it differently?
- Could you have reasonably taken any other action?
- Would you have made the same decision at an earlier point in your career?
- Would this incident have turned out differently if you, or someone with your level of skill/experience had not been there?

Listen for: Other possible courses of action, other potential interpretations, expert-novice differences, potential errors.

Goals: What were your specific goals at this time? What were you hoping/intending to accomplish at this point?

Listen for: Critical decisions, cues and their implications, ambiguous cues, strategies, anomalies/violated expectancies.

Figure A.1

Template for conducting a CDM interview.

the activities might be identified with locations where individual work roles might be identified with particular work locations and so on.

It must be kept in mind that no observation is unobtrusive. In arguing for the advantages of so-called unobtrusive observations, people sometimes assert that in the ideal case, the observer should be like a fly on the wall. The analogy is telling, because when someone is concentrating at work, a fly in the room, on the wall or otherwise, can be very distracting. The goal of observations and in situ interviews for CTA is not to capture behavior that has not been influenced in any way by the presence of the researcher, but to capture *authentic* behavior on the part of the worker (Woods 1993). The researcher is far more likely to observe authentic behavior if he or she has to some extent already been accepted into the culture of the work than if he or she tries to be a fly on the wall. Acceptance means that the workers regard the researcher as informed, sincere, and oriented toward helping, and the workers know that the researcher respects them for their experience and skill.

This understanding of observational methods means that the possibilities for observation include merging interviewing and observing, hence our reference to in-situ interviewing. It is important that the analyses be conducted in the work place, since artifacts in the workplace can serve as contextual cues to both the participant and the researcher. The analysis can be conducted with the assistance of an informant participant, who accompanies the researcher and answers questions. Questions can cover a variety of topics, indicated in table A.1.

The roles analysis and locations analysis of in situ interviews, described later, are conducted for the purpose of specifying work patterns at the social level (information-sharing) and at the individual level (cognitive activities). To focus on roles, the participant is asked questions about each of the jobs (duty assignments) that focus on the flow of information, such as, "What information does this person need?" and, "What does s/he do with the information?" To focus on locations, the participant is asked questions about each of the tools (workstations), such as "What tasks are conducted here?" What is the action sequence involved?" and, "What makes the task difficult?"

Workplace, locations, and roles analyses, activities observations, and standard operating procedure (SOP) analyses involve variations on similar probe questions, but the purposes of the analyses can differ, even though all can culminate in decision requirements tables (DRTs) and/or action requirements tables (ARTs).

The Importance of Opportunism

Once a workplace map (see the following section on workspace analysis) has been prepared, every time the researchers visit the workplace they should bring with them a

Table A.1
Example of possible probes and things to look for

Possible Probes

What subtasks or action sequences are involved in this activity?

What information does the practitioner need?

Where does the practitioner get this information?

What does the practitioner do with this information?

What forms have to be completed?

How does the practitioner recover when glitches cause problems?

How does the practitioner do workarounds?

Is each piece of technology a legacy system or a mandated legacy system?

Things to Look For

Multitasking and multiple distractions?

Do procedures require preparation of reports having categories and descriptions with little meaning?

Are the categories and descriptions really relevant to job effectiveness?

Is the environment one in which it is hard to develop skill?

Do task demands match with the equipment?

Are apprentices or trainees overwhelmed? Do they have to work through tons of data?

What circumstances induce conservatism?

Do workers test their own processes so as to learn the extent of their capabilities?

Do workers have leeway to make risky decisions or other opportunities to benefit from learning?

Does any mentoring occur? What are the opportunities or the obstacles?

Is their routine work (such as reporting functions) that must be performed detracting from learning or understanding?

Do tasks or duties force the worker to adopt one or another style or strategy?

---

folder containing copies of the workplace map and also copies of the data collection form being used to record observations of worker activities. At any moment there may be an opportunity to take notes about work patterns (e.g., information sharing) or an observation that suggests leverage points.

Workspace Analysis

The main goal of this analysis is to create a detailed representation of the workspace to inform subsequent analyses of work patterns and work activities. Activities observations, roles analysis, locations analysis, action requirements analysis, and SOP analysis can all rely on the workplace map, and in some circumstances will rely heavily on it.

Workspace analysis consists of creating a detailed map of the workplace. This is not as simple as it sounds because a detailed map, drawn to scale, takes considerable care and attention to detail. The workplace can be photographed to any level of detail, including views of individual workspaces, photographs of physical spaces alone, photographs of individuals at work, and so on. The workplace should be photographed from a variety of perspectives, for example from the entranceways, from each desk or workstation looking toward the other desks, and so on. It is important to note all of the desks, individual worker's workplaces, the locations of cabinets, records, operating manuals, and the like. Using the photographs, a preliminary sketch of the workplace is refined into a map of the workplace, noting any special features that are pertinent to the research goals (for example, locations of information resources, obstacles to communication, and so on).

One never knows beforehand what things in the work place may turn out to be important (or unimportant). A detailed map and set of photographs can be repeatedly referred to throughout the research program and mined for details that are likely to be missed at the time that the photographs and map are made. A simple device such as a paper cutter or a jumbled pile of reference manuals could turn out in a later analysis to be an important clue to aspects of the work.

Here are examples from the weather forecasting project (Hoffman, Coffey, and Ford 2000):

• One of the participant interviews revealed information about problems forecasters were having with a particular software system. Subsequently, the photographs were examined and revealed that the forecasters had applied a relatively large number of stick-on notes at the workstation, confirming the interview statements.
• Interviews with participating expert forecasters revealed that some of them had kept hardcopy files of images and other data regarding tough cases of forecasting that they had encountered. Using the workplace map, the researchers were able to create a second map showing the locations of the information resources, which in turn could be used in the study of work patterns.
• Photographs of the workplace showing the participants at work showed how forecasters provided information to pilots and revealed obstacles to communication and ways in which the workspace layout was actually a hindrance. The workplace was subsequently reconfigured.

The workplace map from the weather project is presented in figure A.2. We present this figure to illustrate the level of detail that can be involved.

Related to the importance of subtle indicators and the importance of conducting a photographic survey, it may also be advisable to take photographs of the workplace throughout the course of the investigation, perhaps even at planned intervals. The

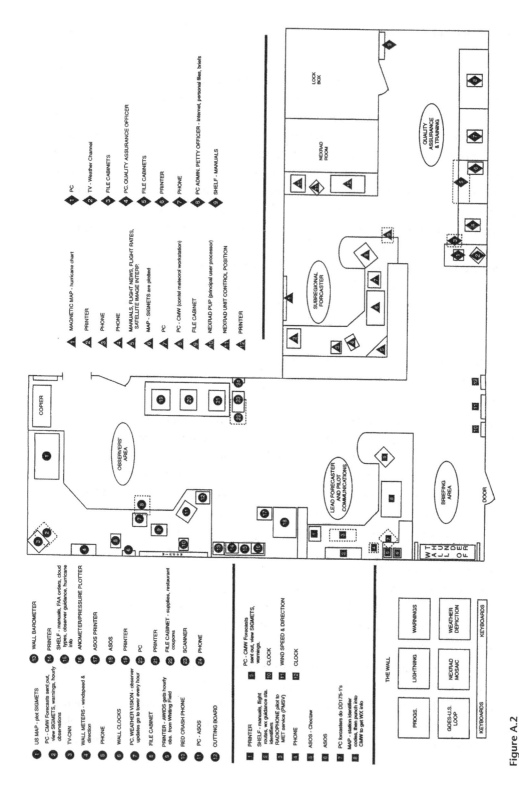

Figure A.2
An example workplace map.

Observer	
Informant(s) (if any)	
Date	

Observation Record

Time	Actor	Activities

Figure A.3
Template for activities observations (ART).

complex sociotechnical workplace is always a moving target, and changes made in the workplace can be important clues to the nature of the work. For instance, in the course of conducting the weather forecasting case study, the main forecasting workstation and its software were changed out completely.

Activities Observations

This simple and deceptively straightforward protocol is for observing people at work. What is not so simple or straightforward is deciding on what to observe and how to record the observations. This necessarily involves some sort of reasoned, specific, and functional means or categorizing or labeling work activities. Like the workspace analysis, this procedure does not involve interviewing. This distinguishes workspace analysis and activities observations from the in situ interview procedures, which combine an element of observing with an element of interviewing. A data collection form for activities observations would be a simple matrix, with columns for time, actor, and activities (see figure A.3).

In Situ Locations Analysis

This analysis also uses a simple matrix data collection form and looks at the work from the perspective of individual locations (desks, cubicles, and so forth) within the total workplace. A participant and the researcher go through the workplace one work location at a time, and the researchers ask questions about it: Who works here? What

Researcher/Observer:
Participant/Informant:
Date:

Instructions for the Participant:
We would like to go through the workplace one work location at a time, and ask you some questions about it. This may take some time and we need not rush through it in a single session. Some of our questions may require a bit of thought. Feel free to take your time, and of course ask any questions of us that come to mind.

Work Space (desk, workstation):
Location (see Workspace Map):
Work Space Layout (Researcher sketches or photographs):

Who works here?

Name	Notes

Tools and Technologies

Tool/Technology	Description, Uses, Notes

Role or activities enacted at the work space

Activity:
What are the goals of this activity?
What skills, knowledge, and experience are needed for successful accomplishment of this activity?
What information is needed for successful accomplishment of this activity?
What about this workspace makes the goal easy to achieve?
What about the workspace makes the goal difficult to achieve?
Kluges and work-arounds:

Figure A.4
Template for locations analysis.

Researcher/Observer:
Participant/Informant:
Date:

Instructions for the Participant
We would like to go through the workplace, one job assignment at a time, and ask you some questions about it. This may take some time and we need not rush through it in a single session. Some of our questions may require a bit of thought. Feel free to take your time, and of course ask any questions of us that come to mind.

Role (job, post, or duty assignment):
Goals:
Tasks:
Needed skills, knowledge, and experience:
What makes the job easy?
What makes the job difficult?

Figure A.5
Template for roles analysis.

tools (software) do they use? What roles/jobs/activities are conducted here? A template for Locations Analysis is provided in figure A.4.

In Situ Roles Analysis

The in situ analysis of roles (jobs, duty assignments, and so on) can be aided by keeping on hand the finished workplace map, but even more important is to conduct the analysis in the workplace rather than outside of it. In collaboration with the researcher, the practitioner goes through the workplace, one job assignment at a time, and is asked some questions about it. Thus, roles analysis has some of the features of observational methods and some of the aspects of an interview. For the roles analysis, the participant is asked about goals, tasks, needed knowledge, technology, and what makes roles easy or difficult (see figure A.5).

In Situ Action Requirements Analysis

This in situ procedure involves an observation or interview about work patterns, resulting in an action requirements table (ART) (figure A.6). For each task goal or activity, the

Interviewer:	Task Designation or Identifying Goal:
Participant:	
Date:	

What is the action sequence?
What cognitive activities are involved in this task/activity?
In what ways can the activity be difficult? What about the support or information depiction makes the action sequence difficult?
What are the informational cues? How are they depicted?
What is the technology or aid, and how does it help? What is good or useful about it?
Are there any work-arounds?
Are there any local "kludges" to compensate for workplace or technology deficiencies?
What kinds of errors can be made?
What kinds of additional aids might be useful?

Figure A.6
Actions Requirements Table.

participant is asked to describe the action sequence, support and tools, and information needs. The researcher's focus is also on issues of usability and usefulness. An ART is an identification and codification of the important activities that are involved in performing a particular task. In addition, the table captures the dynamic flow of activity. In the analysis, one ART is completed for each task or goal. The ART specifies:

- The action sequence involved in the task,
- The equipment, tools, forms, and the like that are used to conduct the task,
- A specification of the information the person needs in order to conduct the task,
- Notes about what is good and useful about the support,
- Notes about ways in which the support makes the task unnecessarily difficult or awkward or requires the creation of workarounds.

As a consequence of these depictions, the ART is intended to suggest new approaches to display design, workspace design, and work patterns design.

In Situ SOP Analysis

One of the first steps in a CTA is preparation, in which the analysts familiarize themselves with the domain. Ordinarily, SOP documents would be a prime source of information for preparation. Cognitive task analysis is sometime conducted solely by means of studying documents. However, it is not unheard of for SOP documents to be a poor reflection of actual practice, for a variety of reasons. Workers may rely on shortcuts and workarounds that they have discovered or created. Standard operating procedures may specify (sub)procedures that are not regarded as necessary. SOPs may simply be ignored, and the reasons for this might be informative. An SOP might be outdated (given that task requirements evolve) or ill specified. For these and other reasons, the analysis of SOP documents can be a window to the "true work."

The analysis of SOP documents depends on the cooperation of a highly experienced domain practitioner who is willing to talk openly and candidly. The participant and the researcher discuss each SOP at the workplace location where the procedure is conducted. For each SOP, the participant is asked the following kinds of questions: Can you briefly describe what this procedure is really about? What are its goals and purposes? What is the basic action sequence? Who does this procedure? How often is the procedure conducted? What is good or easy about the action sequence? When you conduct the sequence, do you really do it in a way that departs from the specifications in the SOP? Are there shortcuts? Do you use a "cheat sheet?" How often have you actually referred to the SOP document for guidance?

When probed about the deficiencies of a certain tool or workstation, the practitioner's initial response to the probe may be meditative silence. Only with some patience and reprobing will deficiencies be mentioned. Often, the practitioner will have discovered deficiencies but will not have explicitly thought about them. Reprobing can focus on makework, workarounds, action sequence alternatives, alternative displays, changes to the tool, and so forth. It is critical to ask about each piece of equipment and technology—whether it is legacy, how it is used, what is good about it, whether its use is mandated, and so on.

While in situ analysis of individual SOPs can be brief, the interviews have to be conducted in two waves, one to draft tables of the results and the second to flesh them out and fine-tune them. Another common circumstance is when the domain includes workers with special areas of proficiency or responsibility. Groups of tasks (and the corresponding SOPs) may be unfamiliar to any one individual. In such cases, SOP interviews will need to be conducted with more than one individual.

Notetaking

It is not always obvious, especially to new CTA researchers, how to create an accurate, complete, and useful data record from interviews or observations. Handwritten notes, typed notes, audio or video recordings—which is necessary? Which is preferable? Our experience is that none of these sufficiently captures an interview or observation to the point of constituting a stand-alone data record.

New CTA researchers might worry about whether they capture everything said in an interview. It is no accident that we recommend two interviewers, both of whom take notes. But we do not count on notes hastily scribbled during an interview to be the data record. Rather, after each interview session we create a complete (and usually electronic) file of the notes as soon as possible. By doing this, the researchers can finish sentences and thoughts, add notes about their impressions or thoughts during the interview session, and generally create a more complete data record.

Videotaping has obvious advantages in terms of data capture, but it also raises issues with confidentiality, security, personnel, written permission, and a high degree of intrusiveness. On top of these practical roadblocks is the tedious and time-consuming task of reviewing and analyzing the video.

As with video, the review, transcription, and analysis of audio data records can be tedious and time-consuming. Both audio and video can be useful supplements to researchers' notes, but the time required to review and correct an entire transcript to ensure accuracy of domain language and acronyms can also be significant. In addition, a transcript gives you the words, but leaves everything else out. Tone of voice, body language, and subtle nuances, facial expressions—all nonverbal communication—is lost in a transcript.

In the end, we believe that a text file, created after the interview from handwritten notes and supplemented when necessary with any audio or video data, is the best overall data record. When it's done well, it represents the cognitive perspective and captures the important elements of the interview efficiently.

Here are some pointers for writing up notes:

- **Write them up as soon as possible** after the interview or observation.
- **Try to retain the participant's style.** The primary data record should contain specific comments, details, stories, or examples in their words and images.
- **Retain the order and sequence of the interview.** You may be tempted to tidy it up by putting all the details about a particular element together, even though the interviewee discussed them at different points. Resist that urge. Mirroring the sequence in

which thoughts and ideas emerged is part of capturing the participant's style and perspective.

• **Separate your notes from the data record.** You may have commentary, interpretations, and ideas that occur as you are writing up notes. It is a great idea to capture this thinking while it is fresh, but it should be clearly labeled in the interview record as "interviewer comments" rather than embedded as though it were part of the interview itself. Similarly, your interpretation of the incident, and the interviewee's role within it, should be distinguished from what the interviewee reported. Submerging your interpretation into the data record skews it.

• **Indicate in the notes at appropriate points the question** to which the participant was responding. This does not have to be a verbatim record, but it will be helpful down the road to know what questions were posed, and what was being answered at that point in the process.

There is no getting around the fact that the primary data record you create will reflect the quality of the notes you took in the observation or interview. One of the CTA skills that develops over time is the ability to tune into relevant content and record it in the notes, leaving out what isn't important. Not even two researchers are always able to write down every word that is said in the interview, so developing this filtering skill is important. Taking notes helps you develop an ear for content. It also means that you are not reliant on technology—laptops, video, or audiotapes—to record the interview for you. In the event that you cannot take a laptop into a facility, or the interviewee isn't comfortable being taped, you should have solid notetaking skills that allow you to create a quality data record on your own with just paper and pen.

Notes

Chapter 1

1. See www.snopes.com for additional information and other versions of this story.

Chapter 2

1. Table 2.1, Knowledge Elicitation Categories and Methods, data are from the CTA Resource website (http://www.ctaresource.com). Methods in table 2.1 were assigned to Knowledge Elicitation categories in accord with the CTA Resource designation of a strong or common association between a method and an "attribute" (that is, category).

2. Table 2.2, Survey of Cognitive Engineering Methods and Uses, is from MITRE's Mental Models website (http://mentalmodels.mitre.org/index.htm). MITRE's website is organized around cognitive engineering methodologies, and does not provide a separate listing of analysis or representation tools.

3. Of course, these caveats would hold for all of the knowledge elicitation methods, and for experimental methods as well.

Chapter 5

1. The stories reported in this section are taken from Hoffman, Coffey, and Ford (2000).

2. We define a team as two or more people working together toward a common goal.

Chapter 7

1. Because the two interviewers have different roles in conducting the interview, their notes may contain differences. The "Second" interviewer frequently has a more detailed record of the interview.

2. "Bradys" is shorthand for bradycardia, a sudden drop in heart rate.

3. This process is similar to what occurs in statistically based analysis, where results of a particular set of statistical tests often point the way to additional tests or a different type of statistical procedure.

4. Because many CTA projects involve small numbers of participants, it can be useful to become familiar with nonparametric statistical methods, which do not have the same assumptions about sample size, normalcy of distribution, or variance that parametric techniques require.

5. As we will see, representations can encompass a wide range of text and graphical formats.

6. However, see Militello (2001) for a discussion of representations of expertise.

7. These initial impressions can also provide input for subsequent interview probes.

Chapter 8

1. The entries in table 8.1 are adapted from a similar list compiled by Orasanu and Connolly (1993), who argue that people use different strategies to make decisions in natural settings rather than in the laboratory.

2. The "situational awareness" construct in Endsley's work has many similarities to the sensemaking function described here.

3. The model does describe decision making under the first two of these conditions; the third variable mentioned here, expertise, did affect the frequency of the RPD strategy.

Chapter 9

1. Many of the ideas contained in this chapter were first presented by David Woods at an ONR Conference in 2000 (see Woods et al. 2002).

2. For reviews of the history of cognitive psychology and affiliated disciplines, see Baars (1986), Gardner (1985), Hoffman (1987), and Hoffman and Deffenbacher (1992).

3. Experimental psychology and applied-industrial psychology in Europe were not punctuated by the behaviorist view.

Chapter 10

1. The FDA issued a cautionary note in 1997, after this study had been conducted, warning about the use of Terbutaline through infusion devices to control premature labor.

2. In actuality, many users of word processing tools moved from WordStar to WordPerfect and then to Word.

Chapter 11

1. Statistical comparisons showed a significant difference at $p < .01$.

2. Although we refer to DCD as involving stages, at any time in a project, data collection and representation may be proceeding in parallel.

Chapter 12

1. Gary Klein and Michael McCloskey worked on this training program with John Schmitt, a former Marine.

Chapter 13

1. The research reported in this chapter has been carried out by Klein Associates researchers for a number of commercial sponsors. The particular clients, and the specific products and services examined, are confidential. In the case studies and examples presented, we omitted brand names and altered other identifying characteristics of products. Our purpose is to convey general information about CTA and market research rather than specific information about a particular product or class of products.

Chapter 14

1. Brian Moon's work on cognitive metrics has been a significant influence on our thinking and on this chapter. Brian worked with us on several early drafts of the chapter, and helped to shape many of the ideas expressed here.

2. This is the gap between the work as shaped by the workplace and the work that really needs to be accomplished. See Vicente (1999).

3. Findings reported here were obtained through a subcontract from SAIC for the Joint Forces Intelligence Command Transformation Support Directorate.

References

Ackerman, F., and C. Eden. 2001. Contrasting single user and networked group decision support systems for strategy making. *Group Decision and Negotiation*, 10: 47–66.

Anderson, J. R. 1982. Acquisition of cognitive skill. *Psychological Review*, 89(4): 369–406.

Anderson, J. R., ed. 1981. *Cognitive skills and their acquisition*. Mahwah, NJ: Lawrence Erlbaum & Associates.

Anderson, J. R., and G. Bower. 1973. *Human associative memory*. Washington, DC: Winston.

Annett, J. 1996. Recent developments in hierarchical task analysis. In *Contemporary ergonomics*, edited by S. A. Robertson, 262–268. London: Taylor & Francis.

Atkinson, R. C., and R. M. Shiffrin. 1968. Human memory: A proposed system and its control processes. In *The psychology of learning and motivation: Advances in theory and research*, edited by K. W. Spence and J. T. Spence, 89–195. New York: Academic Press.

Ausubel, D. P. 1963. *The psychology of meaningful verbal learning*. New York: Grune and Stratton.

Ausubel, D. P. 1968. *Educational psychology: A cognitive view*. New York: Holt.

Ausubel, D. P., and J. D. Novak. 1978. *Educational psychology: A cognitive view*. 2nd ed. New York: Holt, Rinehart and Winston.

Baars, B. J. 1986. *The cognitive revolution in psychology*. New York: Guilford.

Bailey, W. A., and D. J. Kay. 1987. Structural analysis of verbal data. In *Human factors in computing systems and graphics interfaces*, edited by J. M. Carroll and P. Tanner, 297–301. London: Academic Press.

Bainbridge, L. 1979. Verbal reports as evidence of the process operator's knowledge. *International Journal of Man-Machine Studies*, 11(4): 411–436.

Balzer, R., and N. Goldman. 1986. Principles of good software specification and their implications for specification languages. In *Software specification techniques*, edited by N. Gehani and A. D. McGettrick, 25–39. Reading, MA: Addison-Wesley.

Barley, S. R., and J. E. Orr. 1997. *Between craft and science: Technical settings in U.S. settings*. London: ILR Press.

Barnes, B. 1974. *Scientific knowledge and sociological theory*. London: Routledge & Kegan Paul.

Baxter, H. C., K. G. Ross, J. K. Phillips, J. Shafer, and J. E. Fowlkes. 2004. Framework for assessment of tactical decision-making simulations. Paper read at Interservice/Industry Training, Simulation, and Education Conference, December 6–9, 2004, Orlando, FL.

Blandford, A., and W. B. L. Wong. 2004. Situation awareness in emergency medical dispatch. *International Journal of Human-Computer Studies*, 61(4): 421–452.

Blumer, H. 1969. *Symbolic interactionism: Perspective and method*. Englewood Cliffs, NJ: Prentice-Hall.

Bonaceto, C., and K. Burns. 2003. Mapping the mountains: A survey of cognitive engineering methods and uses. Abstract in Proceedings of the 6th Conference on Naturalistic Decision Making.

Briggs, G., D. Shamma, and A. J. Cañas. 2001. *Return to Mars 2002* [cited 2005]. Available from http://cmex.coginst.uwf.edu.

Brown, J. S., S. Denning, K. Groh, and L. Prusak. 2004. *Storytelling in organizations: Why storytelling is transforming 21st century organizations and management*. Burlington, MA: Butterworth-Heinemann.

Brown, R., and J. Berko. 1960. Word association and the acquisition of grammar. *Child Development*, 31: 1–14.

Buchanan, B. G., E. A. Feigenbaum, and J. Lederberg. 1971. A heuristic programming study of theory formation in science. In *Proceedings of the Second International Joint Conference on Artificial Intelligence*, edited by D. C. Cooper, 40–50. London: IJCAI.

Buckley, B. W., and L. M. Leslie. 2000. Sudden temperature changes in the Sydney Basin: Climatology and case studies during the Olympic months of September and October. *International Journal of Climatology*, 20: 1533–1541.

Burton, A. M., N. R. Shadbolt, A. P. Hedgecock, and G. Rugg. 1987. A formal evaluation of a knowledge elicitation techniques for expert systems: Domain 1. In *Research and development in expert systems*, edited by D. S. Moralee, 35–46. Cambridge: Cambridge University Press.

Burton, A. M., N. R. Shadbolt, G. Rugg, and A. P. Hedgecock. 1988. A formal evaluation of knowledge elicitation techniques for expert systems: Domain 1. In *Proceedings, First European Workshop on Knowledge Acquisition for Knowledge-Based Systems*. D3.1–21. Reading, England: Reading University.

Buzan, T., and B. Buzan. 1996. *The Mind Map book: How to use radiant thinking to maximize your brain's untapped potential*. New York: Plume.

Cañas, A. J. 1998. *Concept maps: New uses and the underlying technology*. Mountain View, CA: NASA-Ames Research Center.

Cañas, A. J. 1999. *Algunas ideas sobre la educación y las herramientas computacionales necesarias para apoyar su implementación*. Spain: Revista RED, Educación y Formación Profesional a Distancia, Ministry of Education.

Cañas, A. J. 2003. *Concept Maps for Mars: Report to the Human-Centered Computing Program, Intelligent Systems Project*. NASA-Ames Research Center [cited 2005]. Available from http://is.arc.nasa.gov/HCC/tasks/CptMaps.html.

Cañas, A. J., J. Coffey, T. Reichherzer, N. Suri, and R. Carff. 1997. El-Tech: A performance support system with embedded training for electronics technicians. *Proceedings of the Eleventh Florida Artificial Intelligence Research Symposium*, Sanibel Island, FL.

Cañas, A. J., J. W. Coffey, M. J. Carnot, P. Feltovich, R. Hoffman, J. Feltovich, and J. D. Novak. 2003. A summary of literature pertaining to the use of Concept Mapping techniques and technologies for education and performance support. Report prepared for the Chief of Naval Education and Training. Pensacola, FL: Institute for Human and Machine Cognition.

Card, S. K., T. P. Moran, and A. Newell. 1983. *The psychology of human-computer interaction*. Hillsdale, NJ: Lawrence Erlbaum & Associates.

Carroll, J. B. 1953. *The study of language: A survey of linguistics and related disciplines in America*. Cambridge, MA: Harvard University Press.

Chandrasekaran, B., and J. R. Josephson. 2000. Function in device representation. *Engineering with Computers, Special Issue on Computer Aided Engineering*, 16: 162–177.

Chi, M. T. H. 1996. Construction self explanations and scaffolded explanation in tutoring. *Applied Cognitive Psychology*, 10: S33–S49.

Chi, M. T. H., and R. A. Bjork. 1991. Modeling expertise. In *In the mind's eye: Enhancing human performance*, edited by D. Druckman and R. A. Bjork, 57–79. Washington, DC: National Academy Press.

Chi, M. T. H., P. J. Feltovich, and R. Glaser. 1980. Representation of physics knowledge by experts and novices. Technical Report No. 2. Pittsburgh, PA: University of Pittsburgh Learning Research and Development Center.

Chi, M. T. H., P. J. Feltovich, and R. Glaser. 1981. Categorization and representation of physics problems by experts and novices. *Cognitive Science*, 5: 121–152.

Chi, M. T. H., R. Glaser, and M. J. Farr, eds. 1988. *The nature of expertise*. Mahwah, NJ: Lawrence Erlbaum & Associates.

Chomsky, N. 1959. A review of B. F. Skinner's verbal behavior. *Language*, 35(1): 26–58.

Christensen-Szalanski, J. J. J. 1993. A comment on applying experimental findings of cognitive biases to naturalistic environments. In *Decision making in action: Models and methods*, edited by G. A. Klein, J. Orasanu, R. Calderwood, and C. E. Zsambok, 252–264. Norwood, NJ: Ablex.

Chu, R. W., C. M. Mitchell, and P. M. Jones. 1995. Using the operator function model and OFMspert as the basis for an intelligent tutoring system: Towards a tutor-aid paradigm for operators of supervisory control systems. *IEEE Transactions on Systems, Man and Cybernetics*, 25(7): 1054–1075.

Chung, G., E. Baker, and A. Cheak. 2002. *Knowledge mapper authoring system*. Los Angeles: University of California.

Chung, G. K. W. K., H. F. O'Neil, and H. E. Herl. 1999. The use of computer-based collaborative knowledge mapping to measure team processes and team outcomes. *Computers in Human Behavior*, 15: 463–493.

Clancey, W. J. 1997. *Situated cognition: On human knowledge and computer representation*. Cambridge: Cambridge University Press.

Clancey, W. J. 2001. Field science ethnography: Methods for systematic observation on an expedition. *Field Methods*, 13: 223–243.

Claparede, E. 1917. La psychologie de l'intelligence. *Scientia*, 22: 353–368.

Claparede, E. 1934. La genese de l'Hypothese: Etude experimentelle. *Archiv de Psychologie*, 24: 1–154.

Clark, H. H., and S. E. Brennan. 1991. Grounding in communication. In *Perspectives on socially shared cognition*, edited by L. B. Resnick, J. M. Levine, and S. D. Teasley, 127–149. Washington, DC: American Psychological Association.

Cofer, C. N. 1979. Human learning and memory. In *The first century of experimental psychology*, edited by E. Hearts, 232–370. Hillsdale, NJ: Lawrence Erlbaum & Associates.

Coffey, J. W., and M. J. Carnot. 2003. Graphical depictions for knowledge generation and sharing. In *Proceedings of IKS2003, IASTED International Conference on Information and Knowledge Sharing*, edited by W. Chu, 18–23. Anaheim, CA: ACTA Press.

Coffey, J. W., D. Moreman, and J. Dyer. 1999. *Knowledge preservation at NASA-Glenn Research Center*. Cleveland, OH: NASA-Glenn Research Center.

Cohen, G. 1989. *Memory in the real world*. Mahwah, NJ: Lawrence Erlbaum & Associates.

Collins, H. M. 1993. The structure of knowledge. *Social Research*, 60: 95–116.

Collins, H. M. 1997. RAT-tale: Sociology's contribution to understanding human and machine cognition. In *Expertise in context*, edited by P. J. Feltovich, K. M. Ford, and R. R. Hoffman, 293–311. Cambridge, MA: MIT Press.

Cooke, N. J. 1994. Varieties of knowledge elicitation techniques. *International Journal of Human-Computer Studies*, 41: 801–849.

Cooke, N. J., E. Salas, J. A. Cannon-Bowers, and R. J. Stout. 2000. Measuring team knowledge. *Human Factors*, 42(1): 151–173.

Craik, F. I. M., and R. S. Lockhart. 1972. Levels of processing: A framework for memory research. *Journal of Verbal Learning and Verbal Behavior*, 11: 671–684.

Crandall, B., and R. Calderwood. 1989. *Clinical assessment skills of experienced neonatal intensive care nurses*. Fairborn, OH: Klein Associates.

Crandall, B., and K. Getchell-Reiter. 1993. Critical decision method: A technique for eliciting concrete assessment indicators from the "intuition" of NICU nurses. *Advances in Nursing Sciences*, 16(1): 42–51.

Crandall, B., G. Klein, L. Militello, and S. Wolf. 1994. *Tools for applied cognitive task analysis*. Technical Report prepared under Contract No. N66001-94-C-7008 for the Naval Personnel Research and Development Center, San Diego, CA. Fairborn, OH: Klein Associates.

Crandall, B. W., M. Kyne, L. Militello, and G. A. Klein. 1992. *Describing expertise in one-on-one instruction*. Fairborn, OH: Klein Associates.

Crandall, B., M. McCloskey, C. Adams, and G. Klein. 1996. *Problem solving in mediation: A cognitive study*. Fairborn, OH: Klein Associates.

Crandall, B., R. M. Pliske, and C. E. Zsambok. 1999. *On-the-Job training: A review of the literature*. Fairborn, OH: Klein Associates.

Creswell, J. W. 2003. *Research design*. Thousand Oaks, CA: Sage.

Cross, N., H. Christiaans, and K. Dorst, eds. 1996. *Analyzing design activity*. Chichester, England: John Wiley and Sons.

Cullen, J., and A. Bryman. 1988. The knowledge acquisition bottleneck: Time for reassessment? *Expert Systems*, 5: 216–225.

de Groot, A. D. 1946/1978. *Thought and choice in chess*. New York: Mouton.

De Keyser, V. 1992. Why field studies? In *Design for manufacturability: A systems approach to concurrent engineering and ergonomics*, edited by M. G. Helander and N. Nagamachi, 305–316. London: Taylor & Francis.

De Keyser, V., F. Decortis, and V. Van Daele. 1988. The approach of Francophone ergonomy: Studying new technologies. In *The meaning of work and technological options*, edited by V. D. Keyser, T. Qvale, B. Wilpert, and A. Ruiz-Quintanilla, 147–163. Chichester, England: Wiley & Sons.

Dekker, S., and D. D. Woods. 1999. Extracting data from the future: Assessment and certification of envisioned systems. In *Coping with computers in the cockpit*, edited by S. Dekker and E. Hollnagel, 7–24. Aldershot, England: Ashgate.

Dekker, S. W. A. 2003. Illusions of explanation: Essay on error classification. *The International Journal of Aviation Psychology*, 13(2): 95–106.

Dekker, S. W. A., J. M. Nyce, and R. R. Hoffman. 2003. From contextual inquiry to designable futures: What do we need to get there? *IEEE Intelligent Systems*, 74–77.

Denning, S. 2005. *The leader's guide to storytelling: mastering the art and discipline of business narrative*. San Francisco, CA: Jossey Bass.

Dodson, D. C. 1989. Interaction with knowledge systems through connection diagrams: Please adjust your diagrams. In *Research and development in expert systems*, 33–46. Cambridge: Cambridge University Press.

Dorsey, D. W., G. E. Campbell, L. L. Foster, and D. E. Miles. 1999. Assessing knowledge structures: Relations with experience and posttraining performance. *Human Performance*, 12: 31–57.

Dugger, M., C. Parker, J. Winters, and J. Lackie. 1999. *Interactions between systems engineering and human engineering*. Office of Naval Research, SC-21 Science and Technology Manning Affordability Initiative [cited 2005]. Available from http://www.manningaffordability.com/s&tweb/PUBS/SE_HE/SE_HE_Inter.pdf.

Duncker, K. 1945. On problem solving. *Psychological Monographs*, 58(5).

Ebright, P. R., E. S. Patterson, B. A. Chalko, and M. L. Render. 2003. Understanding the complexity of registered nurse work in acute care settings. *Journal of Nursing Administration*, 33(12): 630–638.

Edwards, J., and K. Fraser. 1983. Concept Maps as reflectors of conceptual understanding. *Research in Science Education*, 13: 19–26.

Eggleston, R. G., M. J. Young, and R. D. Whitaker. 2000. Work-centered support system technology: A new interface client technology for the battlespace infosphere. In *Proceedings of IEEE, NAECON*, Dayton, OH.

Ehn, P. 1988. *Work-oriented design of computer artifacts*. Stockholm: Arbeslivscentrum.

Elm, W. C., S. S. Potter, J. W. Gualtieri, J. R. Easter, and E. M. Roth. 2003. Applied cognitive work analysis: A pragmatic methodology for designing revolutionary cognitive affordances. In *Handbook of cognitive task design*, edited by E. Hollnagel, 357–382. Mahwah, NJ: Lawrence Erlbaum & Associates.

Endsley, M. R. 1988a. Design and evaluation for situation awareness enhancement. In *Proceedings of the Human Factors Society 32nd Annual Meeting*, 97–101. Santa Monica, CA: Human Factors Society.

Endsley, M. R. 1988b. Situation awareness global assessment technique (SAGAT). *Proceedings of the National Aerospace and Electronics Conference (NAECON)*: 789–795.

Endsley, M. R. 1995a. Measurement of situation awareness in dynamic systems. *Human Factors*, 37(1): 65–84.

Endsley, M. R. 1995b. Situation awareness and the cognitive management of complex systems. *Human Factors*, 37(1): 85–104.

Endsley, M. R. 1995c. Toward a theory of situation awareness in dynamic systems. *Human Factors*, 37(1): 32–64.

Endsley, M. R. 1997. The role of situation awareness in naturalistic decision making. In *Naturalistic decision making*, edited by C. Zsambok and G. Klein, 269–283. Mahwah, NJ: Lawrence Erlbaum & Associates.

Endsley, M. R., B. Bolte, and D. G. Jones. 2003. *Designing for situation awareness: An approach to human-centered design*. London: Taylor & Francis.

Endsley, M. R., and D. J. Garland. 2000. *Situation awareness: Analysis and measurement*. Mahwah, NJ: Lawrence Erlbaum & Associates.

Endsley, M. R., and W. M. Jones. 2001. A model of inter- and intrateam situational awareness: Implications for design, training, and measurement. In *New trends in cooperative activities: Understanding system dynamics in complex environments*, edited by M. McNeese, E. Salas, and M. Endsley, 46–67. Santa Monica, CA: Human Factors & Ergonomics Society.

Endsley, M. R., S. J. Selcon, T. D. Hardiman, and D. G. Croft. 1998. *A comparative evaluation of SAGAT and SART for evaluations of situation awareness*. In *Proceedings of the Human Factors and Ergonomics Society Annual Meeting*, 82–86. Santa Monica, CA: Human Factors and Ergonomics Society.

Ericsson, K. A. 1996. The acquisition of expert performance: An introduction to some of the issues. In *The road to excellence: The acquisition of expert performance in the arts and sciences, sports, and games*, edited by K. A. Ericsson, 1–50. Mahwah, NJ: Lawrence Erlbaum & Associates.

Ericsson, K. A., R. Krampe, and C. Tesch-Römer. 1993. The role of deliberate practice in the acquisition of expert performance. *Psychological Review*, 100(3): 363–406.

Ericsson, K. A., and H. A. Simon. 1984. *Protocol analysis: Verbal reports as data*. Cambridge, MA: MIT Press.

Ericsson, K. A., and H. A. Simon. 1993. *Protocol analysis: Verbal reports as data*. 2nd ed. Cambridge, MA: MIT Press.

Ericsson, K. A., and J. Smith. 1991. *Toward a general theory of expertise: Prospects and limits*. Cambridge: Cambridge University Press.

Feltovich, P. J., K. M. Ford, and R. R. Hoffman. 1997. *Expertise in context*. Menlo Park, CA: AAAI Press.

Feltovich, P. J., R. R. Hoffman, D. Woods, and A. Roesler. 2004. Keeping it too simple: How the reductive tendency affects cognitive engineering. *IEEE Intelligent Systems*, 19(3): 90–94.

Fisher, K. M. 1990. Semantic networking: The new kid on the block. *Journal of Research in Science Teaching*, 27: 1001–1018.

Fitts, P. M., and M. I. Posner. 1967. *Human performance*. Belmont, CA: Brooks Cole.

Flach, J. M., and C. O. Dominguez. 1995. Use-centered design: Integrating the user, instrument, and goal. *Ergonomics in Design*, 3(3): 19–24.

Flanagan, J. C. 1954. The critical incident technique. *Psychological Bulletin*, 51: 327–358.

Flavell, J. 1963. *The developmental psychology of Jean Piaget.* New York: Van Nostrand.

Fleck, J., and R. Williams, eds. 1996. *Exploring expertise.* Edinburgh: University of Edinburgh Press.

Flin, R., E. Salas, M. Strub, and L. Martin, eds. 1997. *Decision making under stress: Emerging themes and application.* Aldershot, England: Ashgate.

Ford, K. M., A. Cañas, J. Jones, H. Stahl, J. Novak, and J. Adams-Webber. 1991. ICONKAT: An integrated constructivist knowledge acquisition tool. *Knowledge Acquisition,* 3: 215–236.

Ford, K. M., J. W. Coffey, A. Cañas, E. J. Andrews, and C. W. Turne. 1996. Diagnosis and explanation by a nuclear cardiology expert system. *International Journal of Expert Systems,* 9: 499–506.

Frank, A. W. 2002. Why study people's stories: The dialogical ethics of narrative analysis. *International Journal of Qualitative Methods,* 1(1), Article 6 [cited 2005]. Available from http://www.ualberta.ca/ijqm.

Gagné, R. M. 1968. Learning hierarchies. *Educational Psychologist,* 6: 1–9.

Gagné, R. M. 1974. Task analysis—its relation to content analysis. *Educational Psychologist,* 11: 11–18.

Galegher, J., R. Kraut, and C. Egido, eds. 1990. *Intellectual teamwork: Social and technical bases of cooperative work.* Mahwah, NJ: Lawrence Erlbaum & Associates.

Galegher, J., and R. E. Kraut. 1990. Technology for intellectual teamwork: Perspectives on research and design. In *Intellectual teamwork: Social and technological foundations of cooperative work,* edited by J. Galegher, R. E. Kraut, and C. Egido, 1–21. Mahwah, NJ: Lawrence Erlbaum & Associates.

Gardner, H. 1985. *The mind's new science: A history of the cognitive revolution.* New York: Basic Books.

Gentner, D., and A. L. Stevens, eds. 1983. *Mental models.* Mahwah, NJ: Lawrence Erlbaum & Associates.

Gibson, J. J. 1957. Survival in a world of probably objects [Review of perception and the representative design of psychological experiments]. *Contemporary Psychology,* 2: 33–35.

Glaser, B. G., and A. L. Strauss. 1967. *The discovery of grounded theory: Strategies for qualitative research.* Chicago: Aldine.

Glaser, R. 1976a. Cognitive psychology and instructional design. In *Cognition and instruction,* edited by D. Klahr, 303–316. Mahwah, NJ: Lawrence Erlbaum & Associates.

Glaser, R. 1976b. Cognitive psychology and instructional design. In *Cognition and instruction,* edited by D. Klahr, 303–315. Hillsdale, NJ: Lawrence Erlbaum & Associates.

Glaser, R. 1984. Education and knowledge: The role of thinking. *American Psychologist,* 39: 93–104.

Glaser, R. 1987. Thoughts on expertise. In *Cognitive functioning and social structure over the life course,* edited by C. Schooler and W. Schaie, 81–94. Norwood, NJ: Ablex.

Glaser, R., A. Lesgold, S. Lajoie, R. Eastman, L. Greenberg, D. Logan, M. Magne, A. Weiner, R. Wolf, and L. Yengo. 1985. Cognitive task analysis to enhance technical skills trainng and assessment. Pittsburgh, PA: Learning Research and Development Center, University of Pittsburgh.

Gonzalez, C., O. Juarez, and J. Graham. 2004. Cognitive and computational models as tools to improve situation awareness. *Proceedings of the 48th Annual Meeting of the Human Factors and Ergonomics Society.* New Orleans, LA, September 20–24.

Goodson, J. L., and C. F. Schmidt. 1990. The design of cooperative person-machine problem-solving systems. In *Cognition, computing and cooperation*, edited by S. Robertson, W. Zachary, and J. Black, 187–223. Norwood, NJ: Ablex.

Gordon, S. E. 1992. Implications of cognitive theory for knowledge acquisition. In *The psychology of expertise: Cognitive research and empirical AI*, edited by R. R. Hoffman, 99–120. New York: Springer-Verlag.

Gordon, S. E., and R. T. Gill. 1997. Cognitive task analysis. In *Naturalistic decision making*, edited by C. E. Zsambok and G. Klein, 131–140. Mahwah, NJ: Lawrence Erlbaum & Associates.

Gordon, S. E., K. A. Schmierer, and R. T. Gill. 1993. Conceptual graph analysis: Knowledge acquisition for instructional system design. *Human Factors*, 35(3): 459–481.

Gordon, T., and R. Zemke. 2000. The attack on ISD. *Training*, 37: 42–53.

Graesser, A. C., and S. E. Gordon. 1991. Question answering and the organization of world knowledge. In *Essays in honor of George Mandler*, edited by G. Craik, A. Ortony, and W. Kessen, 227–243. Mahwah, NJ: Lawrence Erlbaum & Associates.

Greeno, J. G. 1978. Understanding and procedural knowledge in mathematics instruction. *Educational Psychologist*, 12: 262–283.

Grover, M. D. 1983. A pragmatic knowledge acquisition methodology. *Proceedings of the 8th International Joint Conference on Artificial Intelligence*, 436–438.

Hammersley, M. 1992. *What's wrong with ethnography: Methodological explorations*. London: Routledge.

Hayes, J. R. 1989. *The complete problem solver*. Hillsdale, NJ: Lawrence Erlbaum & Associates.

Heuer, R. J., Jr. 1999. *Psychology of intelligence analysis*. Washington, DC: Center for the Study of Intelligence, Central Intelligence Agency.

Hoeft, R. M., F. Jentsch, M. E. Harper, A. W. Evans, III, D. G. Berry, C. A. Bowers, and E. Salas. 2002. Structural knowledge assessment with the Team Performance Laboratory's Knowledge Analysis Test Suite (TPL-KATS). *Proceedings of the 46th Annual Human Factors and Ergonomics Society*, 756–760.

Hoffman, R., G. Trafton, and P. Roebber. 2006. *Minding the weather: How expert forecasters reason*. Cambridge, MA: MIT Press.

Hoffman, R. R. 1986. *Procedures for efficiently extracting the knowledge of experts*. Report to the Office of the Deputy for Development Plans, The Strategic Planning Directorate of the Electronic Systems Division, Hanscom AFB, MA. Air Force Office of Scientific Research Contract No. F49260-85-C-0013.

Hoffman, R. R. 1987. The problem of extracting the knowledge of experts from the perspective of experimental psychology. *AI Magazine*, 8: 53–67.

Hoffman, R. R. 1991. Human factors psychology in the support of forecasting: The design of advanced meteorological workstations. *Weather and Forecasting*, 6: 98–110.

Hoffman, R. R. 1997. Human factors in radar meteorology. Presented at the Short Course on Human Factors Applied to Graphical User Interface Design, held at the 28th Conference on Radar Meteorology, American Meteorological Society.

Hoffman, R. R. 1998. Revealing the reasoning and knowledge of expert weather forecasters. Presented at the Fourth International Conference on Naturalistic Decision Making, Warrenton, VA.

Hoffman, R. R., ed. 1992. *The psychology of expertise: Cognitive research and empirical AI*. New York: Springer-Verlag.

Hoffman, R. R., ed. In press. *Expertise out of context: Proceedings of the Sixth International Conference on Naturalistic Decision Making*. Mahwah, NJ: Lawrence Erlbaum & Associates.

Hoffman, R. R., E. L. Cochran, and J. M. Nead. 1990. Cognitive metaphors in the history of experimental psychology. In *Metaphors in the history of psychology*, edited by D. Leary, 173–209. Cambridge: Cambridge University Press.

Hoffman, R. R., J. W. Coffey, and K. M. Ford. 2000. *A case study in the research paradigm of human-centered computing: Local expertise in weather forecasting*. Report on the contract, "Human-Centered System Prototype," National Technology Alliance.

Hoffman, R. R., B. W. Crandall, and N. R. Shadbolt. 1998. Use of the critical decision method to elicit expert knowledge: A case study in cognitive task analysis methodology. *Human Factors*, 40(2): 254–276.

Hoffman, R. R., and K. A. Deffenbacher. 1992. A brief history of applied cognitive psychology. *Applied Cognitive Psychology*, 6: 1–48.

Hoffman, R. R., and K. A. Deffenbacher. 1993. An analysis of the relations of basic and applied science. *Ecological Psychology*, 2(3): 309–315.

Hoffman, R. R., P. J. Feltovich, K. M. Ford, D. D. Woods, G. Klein, and A. Feltovich. 2002. A rose by any other name ... would probably be given an acronym. *IEEE Intelligent Systems*, 17(4): 72–80.

Hoffman, R. R., P. Hayes, K. M. Ford, and P. A. Hancock. 2002. The Triples Rule. *IEEE: Intelligent Systems*, 62–65.

Hoffman, R. R., and R. Hewett. 2001. *The Thailand national knowledge base demonstration project.* Institute for Human and Machine Cognition [cited 2005]. Available from http://www.ihmc.us/research/projects/ThailandKnowledgeBase/.

Hoffman, R. R., G. Klein, and K. R. Laughery. 2002. The state of cognitive systems engineering. *IEEE Intelligent Systems,* 17(1): 73–75.

Hoffman, R. R., and G. Lintern. 2006. Knowledge elicitation. In *Cambridge handbook of expertise and expert performance,* edited by P. Feltovich and R. Hoffman. New York: Cambridge University Press.

Hoffman, R. R., A. Roesler, and B. M. Moon. 2004. What is design in the context of human-centered computing? *IEEE Computer Society,* 89–95.

Hoffman, R. R., and R. J. Senter. 1978. Recent history of psychology: Mnemonic techniques and the psycholinguistic revolution. *The Psychological Record,* 28: 3–15.

Hoffman, R. R., N. R. Shadbolt, A. M. Burton, and G. Klein. 1995. Eliciting knowledge from experts: A methodological analysis. *Organizational Behavior and Human Decision Processes,* 62(2): 129–158.

Hoffman, R. R., and D. D. Woods. 2000. Studying cognitive systems in context: Preface to the Special Section. *Human Factors,* 42(1): 1–7.

Hogarth, R. 2001. *Educating intuition.* Chicago: University of Chicago Press.

Hollnagel, E., G. Mancini, and D. D. Woods, eds. 1986. *Intelligent decision support in process environments.* Berlin: Springer-Verlag.

Hollnagel, E., and D. D. Woods. 1983. Cognitive Systems Engineering: New wine in new bottles. *International Journal of Man-Machine Studies,* 18: 583–600.

Howell, W. C., and N. J. Cooke. 1989. Training the human information processor: A look at cognitive models. In *Training and development in work organizations: Frontiers of industrial and organizational psychology,* edited by I. L. Goldstein, 121–182. San Francisco, CA: Jossey-Bass.

Hutchins, E. 1980. *Culture and inference.* Cambridge, MA: Harvard University Press.

Hutchins, E. 1990. The technology of team navigation. In *Intellectual teamwork: Social and technical bases of cooperative work,* edited by J. Galegher, R. Kraut, and C. Egido, 191–220. Hillsdale, NJ: Lawrence Erlbaum & Associates.

Hutchins, E. 1995a. *Cognition in the wild.* Cambridge, MA: MIT Press.

Hutchins, E. 1995b. How a cockpit remembers its speeds. *Cognitive Science,* 19: 265–288.

Hutton, R. J. B., D. Anastasi, M. L. Thordsen, R. R. Copeland, G. Klein, and D. Serfaty. 1997. *Cognitive function model: Providing design engineers with a model of skilled human decision making.* Contract No. N00178-97-C-3019 for the Naval Surface Warfare Center, Dahlgren, VA. Fairborn, OH: Klein Associates.

Hutton, R. J. B., G. P. Chubb, D. A. Malek, and E. L. Rall. 2003. *CASS study results*. Fairborn, OH: Klein Associates.

Hutton, R. J. B., D. W. Klinger, and B. Crandall. 2003. *The Cognimeter^{SM}: A screening tool for CTA*. Fairborn, OH: Klein Associates.

Hutton, R. J. B., and L. G. Militello. 1996. Applied cognitive task analysis (ACTA): A practitioner's window into skilled decision making. In *Engineering psychology and cognitive ergonomics: Job design and product design*, edited by D. Harris, 17–23. Aldershot, England: Ashgate.

Hutton, R. J. B., L. G. Militello, and T. E. Miller. 1997. Applied cognitive task analysis (ACTA) instructional software: A practitioner's window into skilled decision making. *Proceedings of the Human Factors and Ergonomics Society 41st Annual Meeting*, 2: 896.

Hutton, R. J. B., T. E. Miller, and M. L. Thordsen. 2003. Decision-centered design: Leveraging cognitive task analysis in design. In *Handbook of cognitive task design*, edited by E. Hollnagel, 383–416. Mahwah, NJ: Lawrence Erlbaum & Associates.

Hutton, R. J. B., R. M. Pliske, G. Klein, D. W. Klinger, L. G. Militello, and T. E. Miller. 1998. *Cognitive engineering implications for information dominance*. Wright-Patterson AFB, OH: Armstrong Laboratory [also published as DTIC No. ADB234698, http://www.dtic.mil].

James, W. 1890. *The principles of psychology*. 2 vols. New York: Henry Holt.

Jeffries, R., A. Turner, P. Polson, and M. Atwood. 1981. The processes involved in designing software. In *Cognitive skills and their acquisition*, edited by R. J. Anderson, 225–283. Mahwah, NJ: Lawrence Erlbaum & Associates.

Jenkins, J. J. 1978. Four points to remember: A tetrahedral model of memory experiments. In *Levels of processing and human memory*, edited by L. Cermak and F. Craik, 429–446. Hillsdale, NJ: Lawrence Erlbaum & Associates.

Johnson, P. E., I. Zualkernan, and S. Garber. 1987. Specification of expertise. *International Journal of Man-Machine Studies*, 26: 161–182.

Johnson-Laird, P. N. 1983. *Mental models: Towards a cognitive science of language, inference, and consciousness*. Cambridge, MA: Harvard University Press.

Jonassen, D. H., M. Tessmer, and W. H. Hannum. 1999. *Task analysis methods for instructional design*. Mahwah, NJ: Lawrence Erlbaum & Associates.

Jordan, B., and A. Henderson. 1995. Interaction analysis: Foundations and practice. *The Journal for the Learning Sciences*, 4: 39–103.

Kaempf, G., D. Klinger, and S. Wolf. 1994. *Development of decision-centered interventions for airport security checkpoints*. Contract DTRS-57-93-C-00129 for the U.S. Department of Transportation. Fairborn, OH: Klein Associates.

Kaempf, G. L., S. Wolf, M. L. Thordsen, and G. Klein. 1992. *Decision making in the AEGIS combat information center*. Fairborn, OH: Klein Associates.

Katz, S., A. Lesgold, E. Hughes, D. Peters, G. Eggan, M. Gordin, and L. Greenberg. 1998. Sherlock 2: An intelligent tutoring system built upon the LRDC Tutor Framework. In *Facilitating the development and use of interactive learning environments*, edited by C. P. Bloom and R. B. Loftin, 227–258. New Jersey: Lawrence Erlbaum & Associates.

Kelley, T., and L. Allender. 1996. *A process approach to usability testing for IMPRINT*. Army Research Laboratory; ARL-TR-1171 [cited 2005]. Available from http://mentalmodels.mitre.org/cog_eng/ reference_documents/a%20process%20approach%20to%20usability%20testing%20for%20 IMPRINT.pdf.

Kieras, D. E. 1988. Toward a practical GOMS model methodology for user interface design. In *Handbook for human-computer interaction*, edited by M. Helander, 135–158. Amsterdam: North Holland.

Kintsch, W. 1974. *The representation of meaning in memory*. Hillsdale, NJ: Lawrence Erlbaum & Associates.

Klahr, D. 1976. *Cognition and instruction*. Hillsdale, NJ: Lawrence Erlbaum & Associates.

Klahr, D., and K. Kotovsky, eds. 1989. *Complex information processing: the impact of Herbert A. Simon*. Mahwah, NJ: Lawrence Erlbaum & Associates.

Klein Associates. 1999. Decision skills training: Instructor guide. Fairborn, OH: Klein Associates.

Klein Associates. 2001. IMPACT: Improving performance through applied cognitive training. CD-ROM. Fairborn, OH: Klein Associates.

Klein, G. 1993. *Naturalistic decision making: Implications for design*. Dayton, OH: CSERIAC.

Klein, G. 1998. *Sources of power: How people make decisions*. Cambridge, MA: MIT Press.

Klein, G. 2000. Cognitive task analysis of teams. In *Cognitive task analysis*, edited by J. M. C. Schraagen, S. F. Chipman, and V. J. Shalin, 417–429. Mahwah, NJ: Lawrence Erlbaum & Associates.

Klein, G. 2001. Features of team coordination. In *New trends in cooperative activities: Understanding system dynamics in complex environments*, edited by M. McNeese, M. R. Endsley, and E. Salas, 68–95. Santa Monica, CA: HFES.

Klein, G. 2004. *The power of intuition*. New York: Doubleday.

Klein, G. 2005. The strengths and limitations of teams for detecting problems. Submitted for publication.

Klein, G., and A. A. Armstrong. 2004. Critical decision method. In *Handbook of human factors and ergonomics methods*, edited by N. Stanton, A. Hedge, K. Brookhuis, E. Salas, and H. Hendrick, 1–8, Chapter 35. London: CRC Press.

Klein, G., P. J. Feltovich, J. M. Bradshaw, and D. D. Woods. 2005. Common ground and coordination in joint activity. In *Organizational Simulation*, edited by W. B. Rouse and K. R. Boff, 139–184. New York: John Wiley & Sons.

Klein, G., and L. Militello. 2001. *Some guidelines for conducting a cognitive task analysis.* Edited by E. Salas. Vol. 1, *Advances in human performance and cognitive engineering research.* Greenwich, CT: JAI, Press.

Klein, G., and L. Militello. 2004. The Knowledge Audit as a method for cognitive task analysis. In *How professionals make decisions*, edited by H. Montgomery, R. Lipshitz, and B. Brehmer, 335–342. Mahwah, NJ: Lawrence Erlbaum & Associates.

Klein, G., J. K. Phillips, D. A. Battaglia, S. L. Wiggins, and K. G. Ross. 2002. *FOCUS: A model of sensemaking.* Interim Report—Year 1, prepared under Contract 1435-01-01-CT-31161 (Dept. of the Interior) for the U.S. Army Research Institute for the Behavioral and Social Sciences, Alexandria, VA. Fairborn, OH: Klein Associates.

Klein, G., J. K. Phillips, E. Rall, and D. A. Peluso. In press. A data/frame theory of sensemaking. In *Expertise out of context: Proceedings of the 6th International Conference on Naturalistic Decision Making*, edited by R. R. Hoffman. Mahwah, NJ: Lawrence Erlbaum & Associates.

Klein, G., R. Pliske, S. Wiggins, M. L. Thordsen, S. L. Green, D. Klinger, and D. Serfaty. 1999. *A model of distributed team performance.* Fairborn, OH: Klein Associates.

Klein, G., R. M. Pliske, B. Crandall, and D. Woods. 2005. Problem detection. *Cognition, Technology, and Work*, 7(1): 14–28.

Klein, G., K. G. Ross, B. M. Moon, D. E. Klein, R. R. Hoffman, and E. Hollnagel. 2003. Macrocognition. *IEEE Intelligent Systems*, 18(3): 81–85.

Klein, G., S. Wiggins, and S. L. Green. 2000. *How do talented computer professionals make job selection decisions?* Fairborn, OH: Klein Associates.

Klein, G., S. L. Wiggins, and W. R. Lewis. 2003. Replanning in the Army brigade command post. In *Proceedings of the 2003 CTA Symposium* (CD-ROM).

Klein, G., and S. Wolf. 1995. Decision-centered training. *Proceedings of the Human Factors and Ergonomics Society 39th Annual Meeting*, 2: 1242–1252.

Klein, G., and S. P. Wolf. 1992. *Modeling the option generation process.* Fairborn, OH: Klein Associates.

Klein, G. A. 1989a. Recognition-primed decisions. In *Advances in man-machine systems research*, edited by W. B. Rouse, 5: 47–92. Greenwich, CT: JAI Press. Also published as DTIC No. ADA240659.

Klein, G. A. 1989b. Strategies of decision making. *Military Review*, 56–64.

Klein, G. A. 1992. Using knowledge engineering to preserve corporate memory. In *The psychology of expertise: Cognitive research and empirical AI*, edited by R. R. Hoffman, 170–187. New York: Springer-Verlag.

Klein, G. A., R. Calderwood, and A. Clinton-Cirocco. 1986. Rapid decision making on the fireground. *Proceedings of the Human Factors and Ergonomics Society 30th Annual Meeting*, 1: 576–580.

Klein, G. A., R. Calderwood, and D. MacGregor. 1989. Critical decision method for eliciting knowledge. *IEEE Transactions on Systems, Man, and Cybernetics*, 19(3): 462–472.

Klein, G. A., and R. Hoffman. 1993. Seeing the invisible: Perceptual/cognitive aspects of expertise. In *Cognitive science foundations of instruction*, edited by M. Rabinowitz, 203–226. Mahwah, NJ: Lawrence Erlbaum & Associates.

Klein, G. A., J. Orasanu, R. Calderwood, and C. E. Zsambok, eds. 1993. *Decision making in action: Models and methods*. Norwood, NJ: Ablex.

Klein, G. A., D. D. Woods, and J. Orasanu. 1993. Conclusions. In *Decision making in action: Models and methods*, edited by G. A. Klein, J. Orasanu, R. Calderwood, and C. E. Zsambok, 401–411. Norwood, NJ: Ablex.

Klinger, D. W., and M. G. Gomes. 1993. A cognitive systems engineering application for interface design. *Proceedings of the Human Factors and Ergonomics Society 37th Annual Meeting*, 16–20.

Klinger, D. W., and B. B. Hahn. 2003. *Handbook of team CTA*. Manual developed under prime contract F41624-97-C-6025 from the Human Systems Center, Brooks AFB, TX. Fairborn, OH: Klein Associates.

Klinger, D. W., and G. Klein. 1999. Emergency response organizations: An accident waiting to happen. *Ergonomics in Design*, 7(3): 20–25.

Klinger, D. W., and M. Thordsen. 1998. Team CTA applications and methodologies. *Proceedings of the Human Factors and Ergonomics Society 42nd Annual Meeting*, 1: 206–209.

Knorr-Cetina, K. 1981. *The Manufacture of knowledge: An essay on the constructivist and contextual nature of science*. Oxford: Pergamon.

Knorr-Cetina, K. 1993. Strong constructivism—From a sociologist's point of view: A personal addendum to Sismondo's paper. *Social Studies of Science*, 23: 555–563.

Kolodner, J. 1993. *Case-based reasoning*. San Mateo, CA: Morgan Kaufmann.

Koopman, P., and R. R. Hoffman. 2003. Work-arounds, make-work, and kludges. *IEEE Intelligent Systems*, 18(6): 70–75.

Lashley, K. S. 1951. The problem of serial order in behavior. In *Mechanisms in behavior*, edited by L. A. Jeffress, 112–146. New York: John Wiley.

Latour, B., and S. Woolgar. 1979. *Laboratory life: The construction of scientific facts*. Thousand Oaks, CA: Sage.

Laughery, K. R. 1989. MicroSaint—A tool for modeling human performance in systems. In *Applications of human performance models to system design*, edited by G. R. McMillan, D. Beevis, E. Salas, M. H. Strub, R. Sutton, and L. Van Breda, 219–230. New York, NY: Plenum Press.

Lave, J. 1988. *Cognition in practice*. New York: Cambridge University Press.

Lave, J. 1997. The culture of acquisition and the practice of understanding. In *Situated cognition: Social, semiotic and psychological perspectives*, edited by D. Kirshner and J. A. Whitson, 63–82. Mahwah, NJ: Lawrence Erlbaum & Associates.

Liberman, A. M., P. C. Delatre, and F. S. Cooper. 1952. The role of selected stimulus variables in the perception of unvoiced stop consonants. *American Journal of Psychology*, 65: 497–516.

Lipshitz, R. 1993. Converging themes in the study of decision making in realistic settings. In *Decision making in action: Models and methods*, edited by G. A. Klein, J. Orasanu, R. Calderwood, and C. E. Zsambok, 103–137. Norwood, NJ: Ablex.

Lipshitz, R., G. Klein, J. Orasanu, and E. Salas. 2001. Rejoinder: A welcome dialogue—and the need to continue. *Journal of Behavioral Decision Making*, 14: 385–389.

Loftus, E. F. 1996. *Eyewitness testimony*. Cambridge, MA: Harvard University Press.

Loftus, E. F., and P. Suppes. 1972. Structural variables that determine problem solving difficulty in computer-assisted instruction. *Journal of Educational Psychology*, 63: 531–542.

Lynch, M. 1991. Laboratory space and the technological complex: An investigation of topical contextures. *Science in Context*, 4: 51–78.

Lynch, M. 1993. *Scientific practice and ordinary action*. Cambridge: Cambridge University Press.

Mace, W. M. 1977. James J. Gibson's strategy for perceiving: Ask not what's inside your head but what your head's inside of? In *Perceiving, acting, and knowing*, edited by R. Shaw and J. Bransford, 43–65. Mahwah, NJ: Lawrence Erlbaum & Associates.

Malek, D. A., S. A. Alvidrez, B. M. Moon, and S. Wei. 2004. *The identification and implementation of a cognitive metrics suite for the evaluation and design of tools to enhance warfighter job performance*. Final report prepared under contract number N00178-03-C-1066 for Naval Surface Warfare Center, Dahlgren, VA. Fairborn, OH: Klein Associates.

Markham, K. M., J. J. Mintzes, and M. G. Jones. 1994. The Concept Map as a research and evaluation tool: Further evidence of validity. *Journal of Research in Science Teaching*, 31: 91–101.

Masters, R. 1992. Knowledge, knerves, and know-how: The role of explicit versus explicit knowledge in the breakdown of complex motor skill under pressure. *British Journal of Psychology*, 83: 343–358.

Mattingly, C. 1991. The narrative nature of clinical reasoning. *The American Journal of Occupational Therapy*, 45: 998–1005.

McCarley, J. S., and C. D. Wickens. 2004. *Human factors concerns in UAV flight*. Federal Aviation Administration [cited March 2005]. Available from www.hf.faa.gov/docs/508/docs/uavFY04Planrpt.pdf.

McCloskey, M. J., P. L. Lake, R. M. Pliske, and G. Klein. 1998. Training decision skills for urban warrior squad leaders. Technical Report submitted to SYNETICS Corporation under Contract NSWCDD No. N00178-95-D-1008, King George, VA. Fairborn, OH: Klein Associates.

McDermott, P. L., and B. Crandall. 2000. Uncovering expertise: How cytotechnologists screen pap smears. In *Proceedings of the 5th Conference on Naturalistic Decision Making* (CD-ROM), edited by H. Friman. Stockholm: National Defence College.

McDonald, N., S. Corrigan, and M. Ward. 2002. Well-intentioned people in dysfunctional systems. Keynote Presented at Fifth Workshop on Human Error, Safety and Systems Development, Newcastle, Australia.

McKeithen, K. B., J. S. Reitman, H. H. Rueter, and S. C. Hirtle. 1981. Knowledge organization and skill differences in computer programmers. *Cognitive Psychology*, 13: 307–325.

Means, B., and S. P. Gott. 1988. Cognitive task analysis as a basis for tutor development: Articulating abstract knowledge representations. In *Intelligent tutoring systems: Lessons learned*, edited by J. Psotka, L. D. Massey, and S. A. Mutter, 35–57. Mahwah, NJ: Lawrence Erlbaum & Associates.

Melcher, J., and J. Schooler. 1996. The misremembrance of wines past: Verbal and perceptual expertise differentially mediate verbal over-shadowing of taste memory. *Journal of Memory and Language*, 35: 231–245.

Miles, M. B., and A. M. Huberman. 1994. *Qualitative data analysis: An expanded sourcebook.* 2nd ed. Thousand Oaks, CA: Sage.

Militello, L. G. 2001. Representing expertise. In *Linking expertise and naturalistic decision making*, edited by E. Salas and G. Klein, 245–262. Mahwah, NJ: Lawrence Erlbaum & Associates.

Militello, L. G., and B. Crandall. 1999. Critical incident/critical decision method. In *Task analysis methods for instructional design*, edited by D. H. Jonassen, M. Tessmer, and W. H. Hannum, 181–192. Mahwah, NJ: Lawrence Erlbaum Associates.

Militello, L. G., and R. J. B. Hutton. 1998. Applied Cognitive Task Analysis (ACTA): A practitioner's toolkit for understanding cognitive task demands. *Ergonomics, Special Issue: Task Analysis*, 41(11): 1618–1641.

Militello, L. G., R. J. B. Hutton, and T. E. Miller. 1996. *Applied Cognitive Task Analysis*. Fairborn, OH: Klein Associates.

Militello, L. G., R. J. B. Hutton, R. M. Pliske, B. J. Knight, and G. Klein. 1997. *Applied Cognitive Task Analysis (ACTA) methodology*. Fairborn, OH: Klein Associates. Also published as DTIC No. ADB234698, http://www.dtic.mil.

Militello, L. G., M. M. Kyne, G. Klein, K. Getchell, and M. L. Thordsen. 1999. A synthesized model of team performance. *International Journal of Cognitive Ergonomics*, 3(2): 131–158.

Militello, L. G., M. M. Kyne, G. Klein, K. Getchell-Reiter, and M. L. Thordsen. 1994. Comparing models of team performance. Fairborn, OH: Klein Associates.

Militello, L. G., E. L. Rall, A. A. Armstrong, and S. L. Green. 2002. ARLADA CDM-Adaptations Year 1 Report. Fairborn, OH: Klein Associates.

Miller, G. A. 1979. *A very personal history.* Cambridge, MA: MIT Center for Cognitive Science.

Miller, G. A., E. Galanter, and K. H. Pribram. 1960. *Plans and the structure of behavior.* New York: Henry Holt.

Miller, T. E., A. A. Armstrong, S. L. Wiggins, A. Brockett, A. Hamilton, and L. Schieffer. 2002. *Damage control decision support: Reconstructing shattered situation awareness.* Final Technical Report-Contract No. N00178-00-C-3041 prepared for Naval Surface Warfare Center, Dahlgren, VA. Fairborn, OH: Klein Associates.

Minsky, M. 1963. Steps toward artificial intelligence. In *Computers and thought,* edited by E. A. Feigenbaum and J. Feldman, 406–450. New York: McGraw-Hill.

Mintzes, J., J. Wandersee, and J. D. Novak. 2000. *Assessing science understanding.* San Diego: Academic Press.

Mostert, E., A. Zacharkiewicz, and E. Fossey. 1996. Claiming the illness experience: Using narrative to enhance theoretical understanding. *Australian Occupational Therapy Journal,* 43: 125–132.

Mumaw, R. J., E. M. Roth, K. J. Vicente, and C. M. Burns. 2000. There is more to monitoring a nuclear power plant than meets the eye. *Human Factors,* 42(1): 36–55.

Mumford, M. D., R. A. Schultz, and J. R. van Doom. 2001. Performance in planning: Processes, requirements, and errors. *Review of General Psychology,* 5: 213–240.

Neisser, U., 1967. *Cognitive psychology.* Englewood Cliffs, NJ: Prentice Hall..

Neisser, U., ed. 1982. *Memory observed: Remembering in natural contexts.* San Francisco: W. H. Freeman.

Newell, A. 1968. On the analysis of human problem solving protocols. In *Calcul et formalization dans les sciences de l'homme,* edited by J. C. Gardin and B. Jaulin, 145–185. Paris: Editions du Centre National de la Recherche Scientifique.

Newell, A. 1973. You can't play a game of 20 questions with nature and win. In *Visual information processing,* edited by W. G. Chase, 283–308. New York: Academic Press.

Newell, A., J. C. Shaw, and H. A. Simon. 1958. Elements of a theory of human problem solving. *Psychological Review,* 65: 151–166.

Newell, A., and H. A. Simon. 1972. *Human problem solving.* Englewood Cliffs, NJ: Prentice-Hall.

Nii, P. 1986a. Blackboard systems: The blackboard model of problem solving and the evolution of blackboard architectures. Part One. *AI Magazine,* 7(2): 38–53.

Nii, P. 1986b. Blackboard systems: The blackboard model of problem solving and the evolution of blackboard architectures. Part Two. *AI Magazine,* 7(3): 82–106.

Nisbett, R. E., and T. D. Wilson. 1977. Telling more than we can know: Verbal reports on mental processes. *Psychological Review,* 84(3): 231–259.

Norman, D. A. 1986. Cognitive engineering. In *User centered system design*, edited by D. A. Norman and S. W. Draper, 31–61. Mahwah, NJ: Lawrence Erlbaum & Associates.

Norman, D. A. 1993. *Things that make us smart: Defending human attributes in the age of the machine.* Reading, MA: Addison-Wesley.

Novak, J. D. 1977. *A theory of education.* Ithaca, NY: Cornell University Press.

Novak, J. D. 1990. Concept maps and Vee diagrams: Two metacognitive tools for science and mathematics education. *Instructional Science*, 19: 29–52.

Novak, J. D. 1991. Clarify with concept maps. *The Science Teacher*, 58: 45–49.

Novak, J. D. 1998. *Learning, creating, and using knowledge.* Mahwah, NJ: Lawrence Erlbaum & Associates.

Novak, J. D., and D. B. Gowin. 1984. *Learning how to learn.* Cambridge: Cambridge University Press.

Obradovich, J. H., and D. D. Woods. 1996. Users as designers: How people cope with port HCI design in computer-based medical devices. *Journal of Human Factors and Ergonomics Society*, 38(4): 574–592.

O'Hare, D., M. Wiggins, A. Williams, and W. Wong. 1998. Cognitive task analysis for decision centred design and training. *Ergonomics*, 41: 1698–1718.

Olson, J. R., and K. J. Biolsi. 1991. Techniques for representing expert knowledge. In *Toward a general theory of expertise: Prospects and limits*, edited by K. A. Ericsson and J. Smith, 240–285. Cambridge: Cambridge University Press.

Olson, W. A., and N. B. Sarter. 2001. Management by consent in human-machine systems: When and why it breaks down. *Human Factors*, 43(2): 255–266.

Omodei, M., A. Wearing, and J. McLennan. 1998. Head-mounted video recording: A methodology for studying naturalistic decision making. In *Decision making under stress: Emerging themes and applications*, edited by R. Flin, M. Strub, E. Salas, and L. Martin, 137–146. Aldershot, England: Ashgate.

Orasanu, J., and T. Connolly. 1993. The reinvention of decision making. In *Decision making in action: Models and methods*, edited by G. A. Klein, J. Orasanu, R. Calderwood, and C. E. Zsambok, 3–20. Norwood, NJ: Ablex.

Orasanu, J., U. Fischer, L. K. McDonnel, J. Davison, K. E. Haars, E. Villeda, and C. Van Aken. 1998. How do flight crews detect and prevent errors? Findings from a flight emulation study. In *Proceeding of the 42nd Annual Meeting of the Human Factors and Ergonomics Society*, 191–195. Chicago, IL.

Orr, J. 1985. *Social aspects of expertise.* Palo Alto, CA: Xerox PARC.

Osgood, C. E., and T. A. Sebeok. 1954. *Psycholinguistics: A survey of theory and research problems.* Bloomington, IN: Indiana University Press.

Oskamp, S. 1965. Overconfidence in case study judgments. *Journal of Consulting Psychology*, 29: 261–265.

Paivio, A. 1975. Neomentalism. *Canadian Journal of Psychology*, 29: 263–291.

Patrick, J. 1992. *Training: Research and practice*. San Diego, CA: Academic Press.

Phillips, J. K., M. McCloskey, P. L. McDermott, S. Wiggins, D. A. Battaglia, and G. Klein. 2001. *Decision skills training for small-unit leaders in military operations in urban terrain*. Research Report No. 1776 for Contract DASW01-99-C-0002, U.S. Army Research Institute for the Behavioral and Social Sciences, Alexandria, VA. Fairborn, OH: Klein Associates.

Phillips, J. K., P. L. McDermott, M. Thordsen, M. J. McCloskey, and G. Klein. 1998. *Cognitive requirements for small unit leaders in military operations in urban terrain*. Alexandria, VA: U.S. Army Research Institute for the Behavioral and Social Sciences. Also published as DTIC No. ADA35505, http://www.dtic.mil.

Pliske, R. M., S. L. Green, B. W. Crandall, and C. E. Zsambok. 2000. The collaborative development of expertise (CDE): A training program for mentors. Paper read at Society for Industrial and Organizational Psychology, at New Orleans, LA.

Pliske, R. M., D. Klinger, R. Hutton, B. Crandall, B. Knight, and G. Klein. 1997. *Understanding skilled weather forecasting: Implications for training and the design of forecasting tools*. Technical Report AL/HR-CR-1997-0003. Brooks AFB, TX: U.S. Air Force Armstrong Laboratory.

Pliske, R. M., L. G. Militello, J. K. Phillips, and D. A. Battaglia. 2001. *Evaluating an approach to decision skills training*. Fairborn, OH: Klein Associates.

Pressley, M., and P. Afflerbach. 1995. *Verbal protocols of reading: The nature of constructively responsive reading*. Hillsdale, NJ: Lawrence Erlbaum & Associates.

Rasmussen, J. 1981. Models of mental strategies in process plant diagnosis. In *Human detection and diagnosis of system failures*, edited by J. Rasmussen and W. B. Rouse, 241–258. New York: Plenum Press.

Rasmussen, J. 1985. The role of hierarchical knowledge representation in decision making and system management. *IEEE Transactions on Systems, Man and Cybernetics*, 2(SMC-15): 234–243.

Rasmussen, J. 1986. *Information processing and human-machine interaction: An approach to cognitive engineering*. New York: North Holland.

Rasmussen, J. 1992. The use of field studies for design workstations for integrated manufacturing systems. In *Design for manufacturability: A systems approach to concurrent engineering and ergonomics*, edited by M. G. Helander and N. Nagamachi, 317–338. London: Taylor & Francis.

Rasmussen, J., and M. Lind. 1981. Coping with complexity. In *First annual European conference on human decision-making and manual control*, edited by H. G. Stassen, 70–91. New York: Plenum Press.

Rasmussen, J., A. M. Pejtersen, and L. P. Goodstein. 1994. *Cognitive systems engineering.* New York: John Wiley.

Rasmussen, J., A. M. Pejtersen, and K. Schmidt. 1990. *Taxonomy for cognitive work analysis.* Report RIS-M-2871. Roskilde, Denmark: RIS National Laboratory.

Rasmussen, J., and W. B. Rouse, eds. 1981. *Human detection and diagnosis of system failures.* New York: Plenum Press.

Readinger, W. O. 2004. How they really think: Capturing the context of consumer decision making. *Quirk's,* 54–58.

Readinger, W. O., K. G. Ross, and B. Crandall. 2004. Application of the critical decision method for market research. Paper presented at the Advances in Qualitative Methods Conference, Edmonton, Canada.

Reason, J. 1987. Generic error-modeling system (GEMS): A cognitive framework for locating common human error forms. In *New technology and human error,* edited by J. Rasmussen, K. Duncan, and J. Leplat, 63–83. New York: John Wiley & Sons.

Regoczei, S. B., and G. Hirst. 1992. Knowledge and knowledge acquisition in the computational context. In *The psychology of expertise: Cognitive research and empirical AI,* edited by R. R. Hoffman, 12–28. Mahwah, NJ: Lawrence Erlbaum & Associates.

Rose, M. 2004. *The mind at work: Valuing the intelligence of the American worker.* New York: Viking Penguin.

Ross, K. G., G. Klein, P. Thunholm, J. F. Schmitt, and H. Baxter. 2003. The recognitional planning model: Application for the objective force unit of action (UA). In *Proceedings of the 2003 CTA Symposium* (CD-ROM).

Ross, K. G., A. P. McHugh, D. S. Harris, and R. M. Pliske. 2003. *Assessment of macrocognition in small unit leader training.* Fairborn, OH: Klein Associates.

Ross, K. G., and J. L. Shafer. 2004. *Macrocognitive knowledge representation and design.* Fairborn, OH: Klein Associates.

Ross, K. G., P. Thunholm, M. A. Uehara, A. McHugh, B. Crandall, D. A. Battaglia, G. Klein, and R. Harder. 2003. *Unit of action battle command: Decision-making process, staff organizations, and collaborations.* Report prepared through collaborative participation in the Advanced Decision Architectures Consortium sponsored by the U.S. Army Research Laboratory under the Collaborative Technology Alliance Program, Cooperative Agreement DAAD19-01-2-0009. Fairborn, OH: Klein Associates.

Roth, E. M. 2002. *Field Observation Methods for Cognitive Task Analysis [Web seminar].* CTA Resource [cited March 2005]. Available from http://www.ctaresource.com.

Roth, E. M., E. S. Patterson, and R. J. Mumaw. 2002. Cognitive engineering: Issues in user-centered system design. In *Encyclopedia of Software Engineering,* edited by J. J. Marciniak, 163–179. New York: Wiley-Interscience.

Salas, E., and G. Klein. 2001. Expertise and naturalistic decision making: An overview. In *Linking expertise and naturalistic decision making*, edited by E. Salas and G. Klein, 3–8. Mahwah, NJ: Lawrence Erlbaum & Associates.

Salas, E., C. Prince, D. P. Baker, and L. Shrestha. 1995. Situation awareness in team performance: Implications for measurement and training. *Human Factors and Ergonomics*, 37: 123–136.

Sarter, N., and D. D. Woods. 2000. Team play with a powerful and independent agent: A full mission simulation. *Human Factors*, 42: 390–402.

Sarter, N. B., D. D. Woods, and C. Billings. 1997. Automation surprises. In *Handbook of human factors/ergonomics*, edited by G. Salvendy, 1926–1943. New York: John Wiley & Sons.

Schmitt, J. F. 1995. How we decide. *Marine Corps Gazette*, 79(10): 16–20.

Schmitt, J. F. 1996. Designing good TDGs. *Marine Corps Gazette*, 80(5): 96–97.

Schmitt, J. F., and G. Klein. 1996. Fighting in the fog: Dealing with battlefield uncertainty. *Marine Corps Gazette*, 80(8): 62–69.

Schmitt, J. F., and G. Klein. 1999. How we plan. *Marine Corps Gazette*, 83(10): 18–26.

Schooler, J. W., and T. Y. Engstler-Schooler. 1990. Verbal overshadowing of visual memories: Some things are better left unsaid. *Cognitive Psychology*, 36–71.

Schraagen, J. M. C., S. F. Chipman, and V. J. Shalin, eds. 2000. *Cognitive task analysis*. Mahwah, NJ: Lawrence Erlbaum & Associates.

Schvaneveldt, R. W., D. W. Dearholt, and F. T. Durso. 1988. Graph theoretic foundations of Pathfinder networks. *Computers and Mathematics with Applications*, 15: 337–345.

Schweickert, R., A. M. Burton, N. K. Taylor, E. N. Corlett, N. R. Shadbolt, and A. P. Hedgecock. 1987. Comparing knowledge elicitation techniques: A case study. *Artificial Intelligence Review*, 1: 245–253.

Scribner, S. 1984. Studying working intelligence. In *Everyday cognition: Its development in social context*, edited by B. Rogoff and S. Lave, 9–40. Cambridge, MA: Harvard University Press.

Scribner, S. 1985. Knowledge at work. *Anthropology and Education Quarterly*, 16(3): 199–206.

Shadbolt, N. R., and A. M. Burton. 1990a. Knowledge elicitation. In *Evaluation of human work: Practical ergonomics methodology*, edited by E. N. Wilson and J. R. Corlett, 321–345. London: Taylor and Francis.

Shadbolt, N. R., and A. M. Burton. 1990b. Knowledge elicitation techniques: Some experimental results. In *Readings in knowledge acquisition*, edited by K. L. McGraw and C. R. Westphal, 21–33. New York: Ellis Horwood.

Shalin, V. L., N. D. Geddes, D. Bertram, M. A. Szczepkowski, and D. DuBois. 1997. Expertise in dynamic, physical task domains. In *Expertise in context: Human and machine*, edited by P. J. Feltovich, K. M. Ford, and R. R. Hoffman, 195–217. Menlo Park, CA: AAAI Press.

Shalin, V. L., and C. L. Verdile. 2003. The identification of knowledge content and function in manual labour. *Ergonomics*, 46(7): 695–713.

Shannon, C. E. 1948. A mathematical theory of communication. *Bell System Technical Journal*, 27: 379–423, 623–656.

Shanteau, J. 1992. Competence in experts: The role of task characteristics. *Organizational Behavior and Human Decision Processes*, 53: 252–266.

Shepherd, A. 2000. *Hierarchical Task Analysis*. New York: Taylor & Francis.

Sieck, W. R., G. Klein, D. A. Peluso, J. L. Smith, and D. Harris-Thompson. 2004. FOCUS: A model of sensemaking. Fairborn, OH: Klein Associates.

Silverman, D. 2001. *Interpreting qualitative data: Methods of analyzing talk, text and interaction*. 2nd ed. London: Sage.

Simon, D. P. 1979. A tale of two protocols. In *Cognitive processes in instruction*, edited by J. Lockhead and J. Clement, 119–132. Philadelphia: Franklin Institute.

Simon, H. A. 1973. The structure of ill-structured problems. *Artificial Intelligence*, 4: 181–201.

Sokolowski, J. A. 2003. *Modeling the decision process of a joint task force commander*. Norfork, VA: Old Dominion University.

Sowa, J. F. 1984. *Conceptual structures: Information processing in mind and machine*. Reading, MA: Addison-Wesley.

Staszewski, J. 2004. Models of expertise as blueprints for cognitive engineering: Applications to landmine detection. *Proceedings of the 48th Annual Meeting of the Human Factors and Ergonomics Society*, New Orleans, LA, 20–24 September 2004.

Staszewski, J., and A. Davison. 2000. Mine detection training based on expert skill. In *Detection and remediation technologies for mines and mine-like targets V, Proceedings of Society of Photo-Optical Instrumentation Engineers 14th Annual Meeting*, edited by A. C. Dubey, J. F. Harvey, J. T. Broach, and R. E. Dugan. SPIE, 4038: 90–101.

Stein, E. 1992. A method to identify candidates for knowledge acquisition. *Journal of Management and Information Systems*, 9: 161–178.

Stein, E. 1997. A look at expertise from a social perspective. In *Expertise in context: Human and machine*, edited by P. J. Feltovich, K. M. Ford, and R. R. Hoffman, 181–194. Cambridge, MA: MIT Press.

Sternberg, R. J., and P. A. Frensch, eds. 1991. *Complex problem solving: Principles and mechanisms*. Mahwah, NJ: Lawrence Erlbaum & Associates.

Strauss, A. L., and J. M. Corbin. 1997. *Grounded theory in practice*. Thousand Oaks, CA: Sage.

Suchman, L. 1990. What is human-machine interaction. In *Cognition, computing and cooperation*, edited by S. Robertson, W. Zachary, and J. Black, 25–55. Norwood, NJ: Ablex.

Suchman, L. A. 1987. *Plans and situated actions: The problem of human-machine communication.* Cambridge: Cambridge University Press.

Thordsen, M. 1991. A comparison of two tools for cognitive task analysis: Concept mapping and the critical decision method. *Proceedings of the Human Factors Society 35th Annual Meeting*, 1: 283–285.

Thordsen, M. 1998. Display design for Navy landing signal officers: Supporting decision making under extreme time pressure. In *Proceedings of the Fourth Americas Conference on Information Systems*, edited by E. Hoadley and I. Benbasat, 255–256. Madison, WI: Omnipress.

Thunholm, P. 2003. Military decision making and planning: Towards a new prescriptive model. Ph.D. dissertation, Stockholm University.

Tolcott, M. A., F. F. Marvin, and P. E. Lehner. 1987. *Effects of early decisions on later judgments in an evolving situation.* Technical Report No. 87-10. Falls Church, VA: Decision Science Consortium.

Turing, A. M. 1936. On computable numbers, with an application to the Entscheidungsproblem. *Proceedings of the London Mathematics Society*, Series 2, 42: 230–265.

U.S. Marine Corps. 2003. *Program Manager for Training Systems (PM TRASYS)* [cited 2005]. Available from http://www.marcorsyscom.usmc.mil/trasys/trasysweb.nsf/All/Tactical%20Decision-making%20Simulation%20(TDS)%20Technology.

Umbers, I. G., and P. J. King. 1981. An analysis of human decision-making in cement kiln control and the implications for automation. In *Fuzzy reasoning and its applications*, edited by E. H. Mamdani and B. R. Gaines, 369–381. London: Academic Press.

Underhill, P. 1999. *Why we buy: The science of shopping.* New York: Simon & Schuster.

Vicente, K. J. 1999. *Cognitive work analysis: Towards safe, productive, and healthy computer-based work.* Mahwah, NJ: Lawrence Erlbaum & Associates.

Vicente, K. J. 2002. Ecological interface design: Progress and challenges. *Human Factors*, 33(1): 62–78.

Vicente, K. J., K. Kada-Bekhaled, G. Hillel, A. Cassano, and B. A. Orser. 2003. Programming errors contribute to death from patient-controlled analgesia: Case report and estimate of probability. *Canadian Journal of Anesthesia*, 50: 328–332.

von Neumann, J. 1958. *The computer and the brain.* New Haven, CT: Yale University Press.

Voss, J. F., S. W. Tyler, and L. A. Yengo. 1983. Individual differences in the solving of social science problems. In *Individual differences in cognition*, edited by R. F. Dillon and R. R. Schmeck, 205–232. New York: Academic Press.

Warwick, W., C. Brockett, S. McIlwaine, R. J. B. Hutton, and B. B. Hahn. 2002. Incorporating models of recognition-primed decisions: Progress and lessons learned. Paper read at Simulation Interoperability Workshop, at Orlando, FL.

Warwick, W., and R. J. B. Hutton. In press. Understanding models of the recognition primed decision from both a computational and theoretical perspective. To appear in R. R. Hoffman (ed.). *Expertise out of context* (invited chapter).

Warwick, W., R. J. B. Hutton, S. McIlwaine, and P. McDermott. 2001. Developing a computational model of recognition-primed decision making. Paper read at 10th Conference on Computer-Generated Forces and Behavioral Representation.

Warwick, W., S. McIlwaine, and R. J. B. Hutton. 2002. Developing a computational model of recognition-primed decision making: Progress and lessons learned. Paper read at 11th Conference on Computer-Generated Forces and Behavioral Representation, at Orlando, FL.

Watson, J. B. 1914. *Behavior: An introduction to comparative psychology*. New York: Henry Holt.

Waugh, N. C., and D. A. Norman. 1965. Primary memory. *Psychological Review*, 72: 89–104.

Weick, K. E. 1995. *Sensemaking in organizations*. Thousand Oaks, CA: Sage.

Weiss, D. J., and J. Shanteau. 2003. Empirical assessment of expertise. *Human Factors*, 45(1): 104–116.

Wielinga, B. J., and J. A. Breuker. 1985. Interpretation of verbal data for knowledge acquisition. In *Advances in artificial intelligence*, edited by T. O'Shea, 3–12. Amsterdam: North-Holland.

Wiener, E. L. 1993. *Intervention strategies for the management of human error*. Moffett Field, CA: NASA-Ames Research Center.

Wiener, N. 1948. *Cybernetics of control and communication in the man and the machine*. New York: John Wiley & Sons.

Williams, K. W. 2004. *A summary of unmanned aerial aircraft accident/incident data: Human factors implications*. *Technical report*. Federal Aviation Administration [cited March 2005]. Available from http://www.cami.jccbi.gov/AAM-400A/Abstracts/Tech_Rep.htm.

Williams, R., W. Faulkner, and J. Fleck, eds. 1998. *Exploring expertise issues and perspectives*. London: Macmillan.

Wilson, T. D. 2002. *Strangers to ourselves: Discovering the adaptive unconscious*. Cambridge, MA: Harvard University Press.

Wilson, T. D., and J. W. Schooler. 1991. Thinking too much: Introspection can reduce the quality of preferences and decisions. *Journal of Personality and Social Psychology*, 60(2): 191–192.

Wong, B. L. W. 2004. Critical Decision Method Data Analysis. In *The Handbook of Task Analysis for Human-Computer Interaction*, edited by D. Diaper and N. Stanton, 327–346. Mahwah, NJ: Lawrence Erlbaum & Associates.

Wong, B. L. W., and A. Blandford. 2001. Situation awareness and its implications for human-system interaction. In *Proceedings of the Australian Conference on Computer-Human Interaction*, edited by M. Apperley, 181–186. Perth, Australia: CHISIG, Ergonomonics Society of Australia.

Wong, B. L. W., P. Sallis, and D. O'Hare. 1997. Eliciting information portrayal requirements: Experiences with the critical decision method. In *Proceedings of the HCI '97 Conference on Computers and People XII*, 397–415.

Woodman, M. 1988. Yourdon dataflow diagrams: A tool for disciplined requirements analysis. *Information and Software Technology*, 30(9): 515–533.

Woods, D. D. 1993. Process-tracing methods for the study of cognition outside of the experimental psychology laboratory. In *Decision making in action: Models and methods*, edited by G. Klein, J. Orasanu, R. Calderwood, and C. E. Zsambok, 228–251. Norwood, NJ: Ablex.

Woods, D. D. 1994. Cognitive demands and activities in dynamic fault management: Abduction and disturbance management. In *Human factors of alarm design*, edited by N. Stanton, 63–92. London: Taylor & Francis.

Woods, D. D. 1995. The alarm problem and directed attention in dynamic fault management. *Ergonomics*, 38(11): 2371–2393.

Woods, D. D. 1996. Decomposing automation: Apparent simplicity, real complexity. In *Automation technology and human performance: Theory and applications*, edited by R. Parasuraman and M. Mouloula, 3–17. Mahwah, NJ: Lawrence Erlbaum & Associates.

Woods, D. D. 1998. Designs are hypotheses about how artifacts shape cognition and collaboration. *Ergonomics*, 41: 168–173.

Woods, D. D. 2002. Steering the reverberations of technology change on fields of practice: Laws that govern cognitive work. *Proceedings of the 24th Annual Meeting of the Cognitive Science Society*, 210–218.

Woods, D. D., and E. M. Roth. 1988. Cognitive engineering: Human problem solving with tools. *Human Factors*, 30(4): 416–430.

Woods, D. D., E. M. Roth, and K. B. Bennett. 1990. Explorations in joint human-machine cognitive systems. In *Cognition, computing and cooperation*, edited by S. Robertson, W. Zachary, and J. Black, 123–158. Norwood, NJ: Ablex.

Woods, D. D., and N. Sarter. 2000. Learning from automation surprises and going sour accidents. In *Cognitive engineering in the aviation domain*, edited by N. Sarter and R. Amalberti, 327–354. Hillsdale, NJ: Lawrence Erlbaum & Associates.

Wood, D. D., D. Tinapple, A. Roesler, and M. Feil. 2002. Studying cognitive work in context: facilitating insight at the intersection of people, technology and work. Cognitive Systems Engineering Laboratory, Institute for Ergonomics, The Ohio State University, Columbus OH at url: http://csel.eng.ohio-state.edu/woodscta

Woodworth, R. S. 1938. *Experimental psychology*. New York: Henry Holt.

Zachary, W. W., J. M. Ryder, and J. H. Hicinbothom. 1998. Cognitive task analysis and modeling of complex decision making. In *Decision making under stress: Implications for training and simulation*, edited by J. A. Cannon-Bowers and E. Salas, 315–344. Washington, DC: American Psychology Association.

Zsambok, C. E., B. Crandall, and L. Militello. 1994. *OJT: Models, programs, and related issues.* Fairborn, OH: Klein Associates.

Zsambok, C. E., G. L. Kaempf, B. Crandall, and M. Kyne. 1996. OJT: A cognitive model of prototype training program for OJT providers. Fairborn, OH: Klein Associates. Also published as DTIC No. ADA314546, http://www.dtic.mil.

Zsambok, C. E., and G. Klein, eds. 1997. *Naturalistic decision making.* Mahwah, NJ: Lawrence Erlbaum & Associates.

Index